# RECASTING BIGFOOT

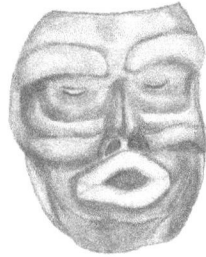

*Books by the author*

# RECASTING BIGFOOT

## DISCOVERING THE TRUTH OF SASQUATCH AMIDST THE HYPE OF BIGFOOT

### GIAN J. QUASAR

*Illustrated*

Library of Congress Cataloging-in-Publication Data

Quasar, Gian J.

Recasting Bigfoot—
Uncovering the truth about Sasquatch amidst the hype of Bigfoot

First Edition

ISBN 978-0-9888505-2-1

Bibliography
1. Bigfoot/cryptozoology    I. Title

TXu 1-861-159

Brodwyn-Moor & Doane

Designed
To evoke the classic age
Of early Sasquatch research

# CONTENTS

# Hype & Hyperbole

From the Yukon to the bluffs of northern California's Klamath Mountains and High Sierras the continent from high above would appear as a tempestuous green sea of furry pines. Like gigantic ocean swells the forests follow the contours of the land, rolling over hills and dipping down into valleys, concealing below as does the sea a world seldom seen. This vast area, this Dark Continent, has been mapped only from these dizzying altitudes. From here rivers appear as silver veins of liquid mercury in the bright sunlight, and those mountain peaks that rise above the forest-line are dangerous, craggy islands swaddled by gusting winds. Intimidating rock precipices are crests of forested waves; their sheer cliffs are streaked with mold and appear to weep mournfully over the dank valleys.

This is as close as most have ever come, and many a seasoned mountain man has never come back. Although we may only guess what happened to them, clues are omnipresent even on a short walk along the fringes of this great unknown. Broken tree stumps form crowns of deadly spikes. Snakes are abundant and lurk in the thickets, and mountain lions have no fear of coming into your camp.

For the most part the dangers seem far from our daily endeavors. We keep within the safe pockets of coastal cities. Then our habitation trickles into outlying small towns. Then it is dabbled into little villages, hamlets, and isolated cabins and logging camps. Our knowledge of the wilderness, even today, is only of its fringes. We follow its life-giving arteries of rushing rivers so far as it is necessary or convenient to hunt and fish, and where they narrow into the veins and capillaries of still creeks and foggy bogs and dells we seldom reach. Even the topography beyond the river banks is shielded from sight, blanketed under the dank silence of the thick underwood, where brambles, thickets, lush ferns and old fallen trees choke from view narrow game paths.

The mystery of the Pacific Northwest therefore is a mystery not of the unexplained but of the unknown. What lies deep within beyond the fringes of the forest lies within the depths of an abyssal sea, one that is harder to access than any ocean, impossible to sound or even consistently to dredge. The sounds we hear are distant echoes: a log rolls, bark cracks, twigs and branches snap; crunches are made by the loping of deer; squirrels scratch their way up and down bark; and there is the chatter of birds and chipmunks, the skipping of the stream, and the hoot of sentinel owls. But there are never blunt sounds. Most everything skulks or is careful in the forest. There is the distant sound of the banjo, the honky-tonk, the logger's buzz saw, the crackle of splitting wood and a swoosh as a tree falls. Man mixes only indistinctly with the quiet sound of the deep forest glades.

This is the true Dark Continent. Whether viewing it from northern California, Oregon, Washington State, Alberta, British Columbia, Yukon and Alaska, it remains the same as it has for thousands of years; it is a quiet, brooding place: foreboding, uninviting and yet alluring.[1]

This is the *terra incognita*, the great unknown; and many have sought to profit by that unknown by selling speculation and potential. The unknown allows us to imagine all sorts of facts and realities behind it. That is the great advantage of mystery— its reality may be anything we wish. Mystery spawns interest. Interest generates hype and hype always gives way to hyperbole, that dreaded world of distorted facts and overblown reactions; that world where synthesis of hearsay and false impression create the reality.

We now stand on the shoulders of 50 years of just such debate on "Bigfoot," an era that saw these dense forests become the center of one of the most fascinating modern legends, a legend that tells us ei-

---

[1] Despite continuing popular impressions, more of America remains wilderness than Africa— 38 % as opposed to Africa's 28 %

ther a primitive "missing link," a bipedal hominid, or "Gentle Giant Ape" thrives within their hidden recesses. Although never having been consistently described, clearly seen, filmed or even proven beyond a doubt to actually exist, the last incarnation— Bigfoot the giant ape —has received popular recognition as a "real species" and even achieved a shaky sort of scientific classification as *Gigantopithecus Americanus*— in other words, a giant ape of America. Though indeed it be Latin, the classification is hardly official. It is made only by those who are self-styled Bigfoot investigators or "crypto-zoologists," the latter those who catalogue things unproven and officially nonexistent.

Unproven species no doubt exist all over the world, with a number of 20th century discoveries such as the African okapi even having previously existed in native legends long before their irrefutable establishment as real and not imaginary or mythological creations. The okapi was not even a composite creation, but was faithfully preserved in native accounts as a real animal, though perhaps sometimes embellished. The controversy about accepting the existence of Bigfoot is that it has undergone remarkably inconsistent and diverse metamorphosis in the last 50 years of general public awareness, changing in shape, size, type of footprint and even its traditional habitat. That in 50 years of verbose theorizing it could transform from missing link or relic hominid and lastly into giant ape is proof of the paucity of sound or consistent evidence.

Prior to this last 50 years of hype, however, there was remarkable consistency. Old newspapers and frontier journals may have reported infrequently upon the matter, but when they did they spoke of frightening encounters with a hairy "wild man" or something "after the fashion of a gorilla and unlike anything ever known." Then later, witnesses who knew what apes looked like confessed that what they had seen could only best be called an "animal human." Except for a few rare occasions, whatever was being seen— man, animal or phantasm— was the size of a human being.

No great legend or public interest came of these encounters because they were so few and far between. Moreover, because they mostly occurred in the remote wilderness of the Pacific Northwest only a fraction of this small number made it to print, and along with this there came the expected ridicule discouraging any future witness from telling his story. To make such stories harder to believe, the Indians remained adamant and vocal that these were the remnants of barbaric tribes which were so different from themselves they could only be called "another species of men." Another "species of men" was something neither science nor religion at the time wanted to hear about. Since apes— the only believable compromise between another "species of men" and common animal— were only proven to exist in the "Old World," such stories were written off as primitive folklore.

It was thus far from hype when J.W. Burns, Canadian teacher and Indian agent for the Saskahaua district of British Columbia, first introduced fellow White Man to the stories of his Chehalis Indians in *MacLean's Magazine* (April, 1929). It took him 3 years of "plodding" to uncover a few Indians who claimed to have had direct contact with the near-extinct wild men of the Saskahaua, the area of British Columbia where the Indians believed the last remnants of the degenerate tribe to still be living. "Their story," he wrote, "is set down here in good faith." There was in Burns' tenor an anticipation of disbelief.

Among the stories of witnesses in his groundbreaking article *Introducing B.C.'s Hairy Giants* was the experience of Indian brave Charley Victor. From his encounter with one Saskahaua klootchman (woman) who could speak and understand the Douglas dialect, Victor lamentably had no choice but to conclude they were somehow related to the Indians. This was a bold statement for any Indian to make. Each Indian tribe had naturally always insisted these degenerates were descendants of enemy tribes and had no relationship to any true Indian.

Victor described the Saskahaua woman in unusual terms for an Indian. "The hairy creature, for that is what it was, walked toward me without the slightest fear. The wild person was a woman. Her face was almost negro black and her long straight hair fell to her waist. In height she would be about six feet, but her chest and shoulders were well above the average in breadth." Like all of her kind, she went about totally naked.

The humanity of these strange and frightening Saskahaua misfits is underscored in the very name given to them in the article— Sasquatch. It's been called an Anglicization, but really it is only a modification on Burns' part of the original Indian pronunciation of Sasqwahachotch— literally "Saskahaua George." In the Chinook Jargon, Brits were Kinchotch— "King George." George was such a common name, it was in some cases assimilated in a casual way for "man" when speaking to Brits; possibly also meaning "foreigner" to the Indians. Thus Sasquatch essentially means Saskahaua man or person. After the term caught on, Sasquatch Man—clearly a redundancy— was also used. (Not surprising since British troops were also known as Kincotch Men.)

Burns' April 1929 article set in motion something he had never intended. Its stories of "ape-like" giant hairy people spread more by the dynamism of the concept rather than by circulation of the article's details, and as such a terrible misperception was created and never dispelled. Specifically, "giant" to the shorter Indian races of the Pacific Northwest, as Burns clarified, had always meant 6 to 6.6 feet tall average but twice the thickness of an ordinary man. Giant to White Man naturally meant something taller. Thanks to Burns' slight Anglicization of the composite Indian name, "Sasquatch" entered White Man's vernacular but thanks to White Man's hearsay it erroneously became an ogre 8 or 9 feet tall.

Only a few years later this error became critical. In 1934, Harrison Hills, 50 miles east of Vancouver, and several Indian settlements on the nearby Chehalis Reservation, were terrorized at night by what the Indians insisted were "Sasquatch men." Jolted from their beds by "eerie wolf-like howling," Whites and Indians alike armed themselves. Morning proved fences were down, livestock released, and storerooms broken into. The rash of events continued over several nights, meriting reports in papers from Vancouver's *Daily Province* to the Fresno *Bee* of California. Finally, the community formed a group of vigilantes to "track the marauders down."

The results, however, were disastrous. Some 20 years later, journalist C.S. Lambert summarized: "However, no specimen of the primitive tribe was captured, and many white people became openly skeptical of the existence of the giants. . .According to Alan Roy Evans, in the *Montréal Standard* ('B.C.'s Hairy Giants'), the Indians are now very sensitive to any imputations cast upon their veracity in this matter. During the 19th century they were ready to tell enquirers all they knew about the Susquatch [sic] men; but today they have become more reserved, and talk only to government agents about the matter.[2] They maintain that the 'Wild Indians' are divided into two tribes, whose rivalry with each other keeps the number down and so prevents them becoming a serious menace to others. . ."

By 1954 when Lambert was summarizing those bizarre events, two decades of Indian silence had only reinforced White Man's last impression that Sasquatches were Indian fairy-tale mountain ogres. After all, how could giants leave no trace whatsoever? The events at Harrison Hills had simply been written off. Thus no one thought it worthy to backtrack and investigate what was perhaps one of the most intriguing events in British Columbian history. And for those who did believe there was truth behind the Indian legends, the van-

---

[2] J.W. Burns is the agent alluded to.

dalizing of Harrison Hills was just the last hurrah of an aboriginal tribe mistakenly thought long extinct.

The tenor of J.W. Burns' last major article— *The Hairy Giants of British Columbia*— also seemed to reflect this same conservatism. This appeared in 1940 in *The Wide World*, the London based magazine for the British Empire. In it he concludes: "Is it possible that primitive 'hairy giants' still inhabit the mountain solitudes of British Columbia? Scientists and others may scoff at the very idea, but many Indians are sincerely convinced that 'Sasquatch'—or at least a few of them— live to this day in the vast, unexplored interior. And, like my Indians, I also believe."

This article introduced a very interesting bit of evidence upon which Burns in some ways relied to justify both his continuing belief and those of his beloved Indians. "Although I have never personally encountered a Sasquatch," wrote Burns, "there is ample proof that hairy giants formerly inhabited the Chehalis district in considerable numbers. Its ancient name— 'The Place of the Wild Men'— was until recently accepted as an echo of primitive superstitions, but the accidental discovery a few years ago of two crude cave dwellings confirmed the Indian legend that the later Troglodytic period of this region was the abode of human beings of huge stature."

Burns was reaching in his conclusion, of course, for abandoned cave dwellings can neither tell us if the inhabitants were giants or hairy.[3] But Burns' leap of faith does not minimize the discovery of those enigmatic caves. They echo legends told over the whole Pacific Northwest that together support his conclusion better than faith.

Almost every tribe, in fact, had told similar stories about a gradual extinction of these gross "wild men" from being tribes to no more than being conclaves living in mountain caves. For example, as early as 1847 the Chinook and Klickitats of Washington State had told artist Paul Kane that Mount St. Helens alone was the remaining abode of the cannibal Skoocooms.[4] The great difficulty in understanding Indian regard for them found its way into Kane's published journal in 1859 in which he described them by the redundancy "race of beings of a different species." The Indians could never qualify such gross violent creatures as humans or as animals. Their reputation

---

[3] C.S. Lambert elaborated: "Expeditions have been organized to track down the Susquatch [sic] men to their lair in the mountains; but the Indians employed to guide these expeditions invariably desert before they reach the danger zone. However, certain large caves have been discovered, with man-made walls of stone inside them, and specially-shaped stones fitted to their mouths, like doors."

[4] Skoocum literally means "strong." Today it can mean well made, even first rate but with a tough edge to it—macho. It is the Chinook trading language. Commonly still in use today for "cool."

made them the embodiment of evil. Accordingly, Kane even referred to them as "Evil Genii." Far from removing the "species" from the real world, it took very real and violent acts to even merit this status. The term Skoocoom was applied to the evilest of predators. Casanov, chief of the Klickitats, was known to have kept a hired assassin. According to Kane, this hated henchman was known by no other name than "Casanov's Skoocoom."

Considering how consistently these stories speak of a tribe of "people" near extinction— the Skoocoom are just one case in point— it is amazing that they have been resurrected today and used to support the notion of a 9-foot tall rogue prehistoric *Gigantopithecus* giant ape roaming all over North America. It is equally amazing that the significance of "two tribes" could ever be overlooked; thus overlooking something very intriguing— that there existed something similar enough to be considered roughly the same thing but different enough for the Indians to draw distinctions.

Curiously, a similar stance was also held by many native peoples in Eurasia concerning the existence there of more than one type of hairy "wild man." Fortunately gigantism, that mistake of Canadian and American Whites, never crept into the Eurasian stories. Therefore descriptions of the "wild men" appeared all the more human in terms, and this led to more serious consideration and study. . .and this led to the formation of the famous Soviet "Snowman Commission" in the 1950s. Amazingly, its members reached the exact opposite conclusion as that held in America. They determined that humans *were* involved.

The Russians had one thing we did not: an authenticated case. The Snowman Commission was able to determine that one of the "Abnauayu," a tribal name applied by the people of the Caucasus Mountains to their 'wild men,' was indeed a human being. A number of the Commission even went so far as to insist that she was a clear case of a living "relic" Neanderthal, a type of cave-dwelling human we thought long extinct. Today, we can say with hindsight's added focus that the descriptions of this "wild woman," who lived in the 19th century, fit that of Charley Victor's savage cannibal Sasquatch woman. While this adds weight to some form of Victor's otherwise improbable story, more concretely it adds weight to the supposition that one of the two Sasquatch "tribes" could indeed have been human, as the Indians themselves always insisted.

The most impressive part of the evidence in Eurasia is that footprints very similar to prehistoric Neanderthals have indeed been found by members of scientific expeditions, some in remote wildernesses. In America no such footprint has been found. Rather the modern American pursuit is documented by nothing more than dozens of different, oftentimes comical feet, all of them now said to support *Gigantopithecus.* Turning to old 19th century frontier newspapers for support doesn't help; it contradicts. Only 2 distinct types of prints are revealed. There is indeed the report of a large footprint with 5 toes; but its humanness can merely be deduced by the fact the hunter does not elaborate on the print or suggests the person making the print would have to wear a huge shoe size. More often than not, old journals describe a 4-toed footprint or a footprint that is unusually long and narrow, tapering toward the heel, the latter two clearly not human.

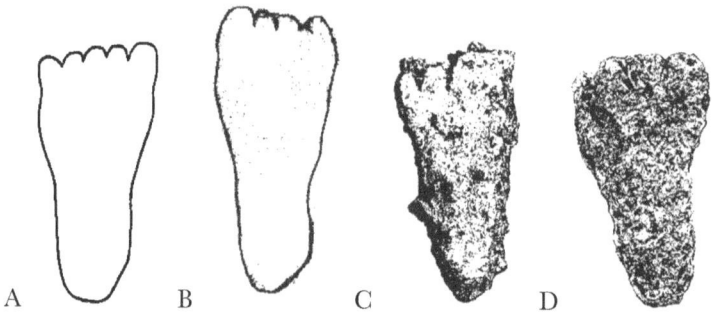

A B C D

    Corroborating these various old Indian and frontiersmen accounts are very rare 20th century discoveries. The first Sasquatch foot ever traced was in 1941 by deputy sheriff Joe Dunn at Ruby Creek, British Columbia (A). It was 16 inches long. Conservational officer R.H. Uchtmann found some similar prints 21 inches in length at The Pass, Manitoba, 1973 (B). They were different only in that they had 4 and not 5 toes. Manitoba, 1988, the Royal Canadian Mounted Police dug these prints out of a muddy road 200 miles north of Winnipeg (C). This is the right foot; also 4-toed. (D) Dunn Lake near Berrière in April 1980, British Columbia; also 4-toed.
    These prints are not those of apes or humans. Nor are they anything in between like the fabled apeman. They suggest rather a *type* of primate our minds have never conceived. One thing they certainly are not. They are not the "Bigfoot" that modern legend has impressed upon us. Yet they are, in fact, the *only* tangible fragment of truth that underlie the Sasquatch and the Skoocoom, those nebulous hairy wild men that in turn underlie "Bigfoot." Ironically, they were hidden not by time but by being merely a few oddities within the vast vault of comical feet that became "Bigfoot."

It was not until October 1958 that we even first hear of the shadowy bearer of big feet. He emerged from the forests in California far from the Saskahaua in the form of enlarged modern human footprints and nothing else. And by a stroke of hyperbole and even greater irony he was declared a "New Sasquatch." A lot has happened since then— myth, exposure, denial. Sadly, all three have only served to cloud the truth of Saskahaua George. The plaster casts of strange funny feet at Bluff Creek, California, have now become a footnote— no pun intended— and with them unfortunately that kernel of truth that underlies the old reports of the "animal humans" has also been archived.

How can one— I, in this case— erase 40 to 50 years of the hype and hyperbole from the popular imagination to reveal again the actual reality? That has been my chief aim ever since beginning to investigate the "world of the unexplained." It takes more than restarting with a blank piece of paper. Popular hearsay survives upon the gasses we breathe. We cannot avoid taking it in, and, unfortunately, unlike the gospeline teaching, it is not harmlessly dispelled in the draught.

It is the purpose of this book to both trace and expose the falsehood of Bigfoot and then reestablish the facts that underpin a truly fascinating reality— that another type of primate exists that is neither man, ape or monkey; and that one tribe of Sasquatches was indeed human. This volume, I think, will prove a fascinating journey through time and history into fact and science, from the past to the present, and in between there is that era of hyperbolic fantasy to give us pause to chuckle.

Quite a few of the "old buffs" of Bigfoot may recognize some of the early reports that will be used. They should not fear that they are about to delve into just some collection of odd stories. This is not a compilation of interviews, a recycling of overused vignettes passed on uncritically, or a dossier of claims. This is a work of analysis, discovery and prediction. The purpose of this book is to weed out fact from fiction and then retrace the steps and follow the true tracks of Sasquatch, not the garish, fake and make-believe. First we must start at the beginning of the evidence. Then we will do the weeding and then we can follow the track of something that even today has only one truly undeniable and tangible characteristic: a big footprint.

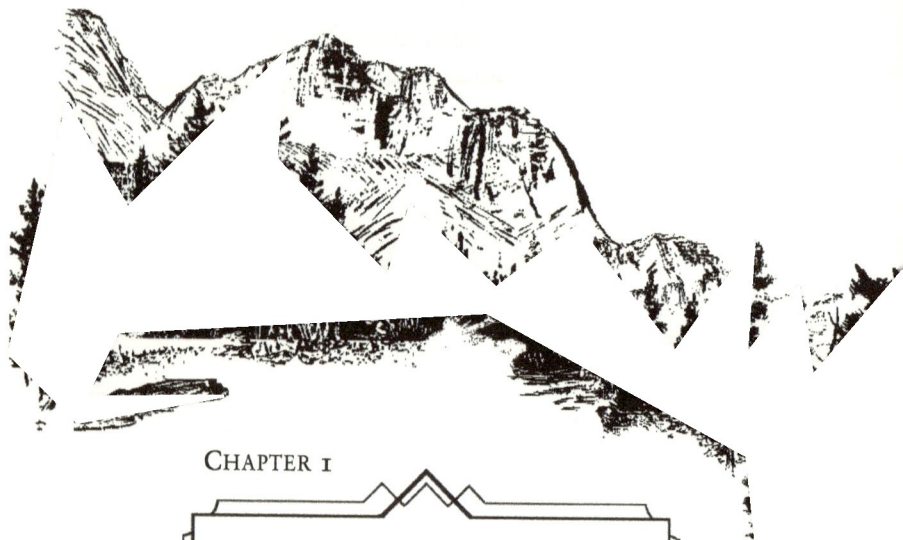

# CHAPTER 1

## Peckatoe Journal

They call this the Dark Continent, this land of dense forests in the Pacific Northwest of the United States and Canada; from northern California to beyond the Peckatoe River in British Columbia; this mountainous land that time forgot, hundreds of thousands of square miles wherein no man has ever stepped foot. There is the sound of the wild in them, of death, of life. . .and eternity too. They have remained unchanged for as long as Man has been civilized, and for just as long they have also proven to be largely impassible. And yet on the border of the glades where Man has settled, or at the base of mountain shoulders, there is also the sound of heavy footsteps within. It is said dogs will not go near the sound, and that even all the casual noise of the forest simultaneously will halt: the crickets stop their chirping, the birds stop their carefree singing; horses will rear and become excited, and a sense of fear or of being watched will overcome even Man.

Apparently dogs can sense the blasted thing. That is the first sign one of 'them' is around; that and a ghastly putrid smell. Dogs will whimper around your legs and put their tail between theirs in fear and submission. Not unless attacked will a dog defend against 'it.'

Beyond the curtain of dense spruce, behind thriving saplings and every conceivable vine and bramble, footsteps have been found, monstrous in size, deeply embedded. There is that foul, sour smell that lingers. There is a momentary chill and then at the nape of the neck there is a tingle. 'It' has been here and now 'it' is gone.

The native Indians have known about 'it' for generations. In northern California the Huppa tribe called them Omah. Further north they are called the Seeahtiks. Other tribes call them Skoocooms, and still others Bukwas or Smy-alik. The Salish Indians of British Columbia call them by what has become the most well-known Indian name: Sasquatch. But even these names don't help identify 'it.' All that these names implied to the Indians were hairy "wild men," cursed men who weren't exactly men and yet who weren't exactly beasts.

When the White Man arrived— the mountain men, fur trappers, Indian traders and explorers— he was tolerantly indulgent of the Indian superstitions. Eventually, however, he too had enigmatic encounters with what he could only describe as "monkey men." Yet again, they weren't exactly monkeys and they weren't exactly men. Throughout the Pacific Northwest its fierceness earned 'it' various names: Creek Devil, Bushman and Mountain Devil.

In his 1892 book *Wilderness Hunter* no less of an authority on big game and the wilderness than Teddy Roosevelt recorded his discussion with an old hunter named Bauman. Roosevelt was a notorious skeptic, and is even the man who coined the term Nature-Faker. Yet even he was impressed enough by Bauman's story to recount it in his book. By the time Roosevelt interviewed Bauman he "had passed his whole life on the frontier" and was now an old, "grizzled and weather beaten" mountain hunter. Whether Roosevelt believed his story or not it is not clear, but he would never have entered it into his book if he had not thought that Bauman was being perfectly sincere. "He must have believed what he said," wrote Roosevelt, "for he could hardly repress a shudder at certain points of the tale."

Teddy Roosevelt takes up the narrative:

When the event occurred Bauman was still a young man, and was trapping with a partner among the mountains dividing the forks of the Salmon from the head of Wisdom River. Not having had much luck, he and his partner determined to go up into a particularly wild and lonely pass through which ran a small stream said to contain many beaver. The pass had an evil reputation because the year before a solitary hunter who had wandered into it was there slain, seemingly by a wild beast, the half-eaten remains being afterward found by some mining prospectors who had passed his camp only the night before.

The memory of this event, however, weighed very lightly with the two trappers, who were as adventurous and hearty as others of their kind . . .They then struck out on foot to the vast, gloomy forest, and in about four hours reached a little open glade where they concluded to camp, as signs of game were plenty.

There was still an hour or two of daylight left, and after building a brush lean-to and throwing down and opening their packs, they started up stream. At dusk they again reached camp.

They were surprised to find that during their absence something, apparently a bear, had visited camp, and had rummaged about among their things, scattering the contents of their packs, and in sheer wantonness destroying their lean-to. The footprints of the beast were quite plain, but at first they paid no particular heed to them, busying themselves with rebuilding the lean-to, laying out their beds and stores, and lighting the fire.

While Bauman was making ready supper, it being already dark, his companion began to examine the tracks more closely, and soon took a brand from the fire to follow them up, where the intruder had walked along a game trail after leaving the camp. . . .Coming back to the fire, he stood by it a minute or two, peering out into the darkness, and suddenly remarked: "Bauman, that bear has been walking on two legs." Bauman laughed at this, but his partner insisted that he was right, and upon again examining the tracks with the torch, they certainly did seem to be made by but two paws, or feet. However, it was too dark to make sure. After discussing whether the footprints could possibly be those of a human being, and coming to the conclusion that they could not be, the two men rolled up in their blankets, and went to sleep under the lean-to.

At midnight Bauman was awakened by some noise, and sat up in his blankets. As he did so his nostrils were struck by a strong, wild-beast odor, and he caught the loom of a great body in the darkness at the mouth of the lean-to. Grasping his rifle, he fired at the vague, threatening shadow, but must have missed, for immediately afterwards he heard the smashing of the underwood as the thing, whatever it was, rushed off into the impenetrable blackness of the forest in the night.

After this the two men slept but little, sitting up by the rekindled fire, but they heard nothing more. In the morning they started out to look at the few traps they had set the previous evening and put out new ones. By an unspoken agreement they kept together all day, and returned to camp towards evening.

On nearing it they saw, hardly to their astonishment, that the lean-to had been again torn down. The visitor of the preceding day had returned, and in wanton malice had tossed about their camp kit and bedding, and destroyed the shanty. The ground was marked up by its tracks, and on leaving the camp it had gone along the soft earth by the brook, where the footprints were as plain as if on snow, and, after careful scrutiny of the trail, it certainly did seem as if, whatever the thing was, it had walked off on but two legs.

The men thoroughly uneasy, gathered a great heap of dead logs, and kept up a roaring fire throughout the night, one or the other sitting on guard most of the time. About midnight the thing came down through the forest opposite, across the brook, and stayed there on the hillside for nearly an hour. They could hear the branches crackle as it moved about, and several times it uttered a harsh, grating, long-drawn moan, a peculiarly sinister sound. Yet it did not venture near the fire.

In the morning the two trappers, after discussing the strange events of the last 36 hours, decided that they would shoulder their packs and leave the valley that afternoon. . . .

All the morning they kept together, picking up trap after trap, each one empty. On first leaving camp they had the disagreeable sensation of being followed. In the dense spruce thickets they occasionally heard a branch snap after they had passed; and now and then there were slight wrestling noises among the small pines to one side of them.

At noon they were back within a couple of miles of camp. In the high bright sunlight their fears seemed absurd to the two armed men, accustomed as they were, through long years of lonely wandering in the wilderness to face every kind of danger from man, brute, or element. There were still three beaver traps to collect from a little pond in a wide ravine near by. Bauman volunteered to gather these and bring them in, while his companion went ahead to camp and made ready the packs.

On reaching the pond Bauman found three beavers in the traps, one of which had been pulled loose and carried into a beaver house. He took several hours in securing and preparing the beaver, and when he started homewards he marked, with some uneasiness, how low the sun was getting. . .

At last he came to the edge of the little glade where the camp lay, and shouted as he approached it, but got no answer. The camp fire had gone out, though the thin blue smoke was still curling upwards. Near it lay the packs wrapped and arranged. At first Bauman could see nobody; nor did he receive an answer to his call. Stepping forward he again shouted, and as he did so his eye fell on the body of his friend, stretched beside the trunk of a great fallen spruce. Rushing towards it the horrified trapper found that the body was still warm, but that the neck was broken, while there were four great fang marks in the throat.

The footprints of the unknown beast-creature, printed deep in the soft soil, told the whole story.

The unfortunate man, having finished his packing, had set down on the spruce log with his face to the fire, and his back to the dense woods, to wait for his companion. . . .It had not eaten the body, but apparently had romped and gambolled round it in uncouth, ferocious glee, occasionally rolling over and over it; and had then fled back into the soundless depths of the woods.

Bauman, utterly unnerved, and believing that the creature with which he had to deal was something either half human or half devil, some great goblin-beast, abandoned everything but his rifle and struck off at speed down the pass, not halting until he reached the beaver meadows where the hobbled ponies were still grazing. Mounting, he rode onwards through the night, until far beyond the reach of pursuit.

Even allowing for some embroidery, the root of Bauman's story may have truth in it, though he may have added some foliage. His account bears uncanny resemblance to an earlier published account, which may be (or the legend behind this published account) the source of his own elaboration.

This account was contained in artist/explorer Paul Kane's wonderful book *The Wanderings of an Artist among the Indians of North America from Canada to Vancouver's Island and Oregon through the Hudson's Bay Company's territory and back again*, which was published in 1859 containing his frontier journals. On March 26, 1847, he recorded:

When we arrived at the mouth of the Kattlepoutal River, twenty-six miles from Vancouver (Washington), I stopped to make a sketch of the volcano, Mt. St. Helens, distant, I suppose, about thirty to forty miles. This mountain has never been visited by either whites or Indians; the latter assert that it is in habited by a race of beings of a different species, who are cannibals, and whom they hold in great dread; they also say that there is a lake at its base with a very extraordinary kind of fish, with a head more resembling that of a bear than any other animal. These superstitions are taken from the statement of a man who, they say, went into the mountain with another, and escaped the fate of his companion, who was eaten by the "skoocooms," or "evil genii." I offered a considerable bribe to any Indian who would accompany me in its exploration but could not find one hardy enough to venture there.

In Bauman's story the bipedality of the assailant was immediately noteworthy. Yet at the same time its footprint was considered nonhuman. This same dichotomy is also implied in the Indian belief that the Skoocoom were "a different species;" manlike enough to be considered cannibalistic for eating humans but not like any men they had ever seen. Apparently, it was with great difficulty that Kane translated and interpreted their impressions, for he will not say another species of men but rather leaves it as the nebulous and redundant "race of beings of a different species."

Another unusual sighting comes from lumber maverick Mike King, who, in 1901, was in the northern forests of Vancouver Island, British Columbia. From this we get yet another impression of the footprint. King was a timber cruiser, lumberman, and a highly successful lumber prospector. He took chances where others wouldn't, largely because he was helped by the local Indians. They had led him to many of his best discoveries merely by knowing the land so well. However, for his latest expedition they refused to go, saying that the territory he was interested in was the domain of the "wild man of the woods."

Big Mike King was not put off by what any white man considered Indian superstition. He set off on his own, and after a couple of days' penetration he stopped on a ridge over a creek to take bearings. It was then that he saw something very unusual. There at the creek below was a brown-haired creature squatting at the bank. Instinctively King raised his gun and aimed at it, thinking he had a brown bear. But when he noticed it was washing roots in the water, which a bear cannot do with its claws, he lowered his gun, squarely amazed, and studied it more carefully. It looked like an animal, but it was behaving in some respects like a human, although he could not see its face clearly. As soon as it noticed King, it took off, to King's surprise running like a man. But there was, of course, several exceptions. King noted: "His arms were peculiarly long and used freely in climbing and brush-running." When he investigated the

[ 21 ]

creek, he noted the footprints. They were human in a way, but "with phenomenally long and spreading toes."

At Myrtle Point in Oregon the agility of an unknown "What is it?" earned him the name "Kangaroo Man." For some 30 years he was reported around the area of the Sixes River mining operations. But it wasn't until December 24, 1900, that two miners' eyewitness accounts finally made the local *Daily Review* of Roseburg. It confirmed that Johnnie McCulloch and William Page saw it come down to the river to drink. When the "animal-man," as they called it, saw them it ran off "with cat-like agility" and was quickly out of sight.

The *Daily Review* synthesized the repetition of sightings of this thing over the preceding 30 years.

The appearance of this animal is almost enough to terrorize the rugged mountaineers themselves. He is described as having the appearance of a man—a very good-looking man— is nine feet in height with low forehead, hair hanging down near his eyes, and is covered with a prolific growth of hair which nature has provided for his protection. Its hands reach almost to the ground and when its tracks were measured its feet were found to be 18 inches in length with five well formed toes. Whether this is a devil, some strange animal or wild man is what Messrs. Page and McCulloch would like to know.

It is hard to imagine how it could be good looking and yet "terrorize" even the rugged mountaineers just by its appearance. Other sightings of this "animal-man" would make it clear that Messrs. Page and McCulloch did not see it clearly or the reporter engaged in some "literary license." By 1904 the "wild man" had been seen several more times, prompting the newspapers to follow up. The following article appeared in the Myrtle Point *Enterprise*, 11 March 1904, and the *Lane County Leader,* 7 April 1904.

At repeated intervals during the past 10 years thrilling stories have come from the rugged Sixes mining district in Coos County, Oregon, near Myrtle Point, regarding the wild man or queer and terrible monster which walks erect and which has been seen by scores of miners and prospectors. . .

The appearance of the 'Wild Man' of the Sixes has thrown some of the miners into a state of excitement and fear. A report says the wild man has been seen three times since the 10th of last month. The first appearance occurred on 'Thompson Flat.' William Ward and a young man by the name of Burlison were sitting by the fire of their cabin one night when they heard something walking around the cabin and which resembled a man walking and when it came to the corner of the cabin it took hold of the corner and gave the building a vigorous shake and kept up a frightful noise all the time— the same that has so many times warned the venturesome miners of the approach of the hairy man and caused them to flee in abject fear.

Mr. Ward walked to the cabin door and could see the monster plainly as it walked away, and took a shot at it with his rifle, but the bullet went wild of its mark. The last appearance of the animal was at the Harrison cabin only a few days ago. Mr. Ward was at the Harrison cabin this time and again figures in the excitement. About five o'clock in the morning the wild man gave the door of the cabin of vigorous shaking which aroused Ward and one of the Harrison boys who took their guns and started in to do the intruder. Ward fired at the man and he answered by sending a 4 lb. rock at Ward's head but his aim was a little too high. He then disappeared in the brush.[5]

Many of the miners avow that the 'wild man' is a reality. They have seen him and know whereof they speak. They say he is something after the fashion of a gorilla and unlike anything else that has ever been known; not only that but he can throw rocks with wondrous force and accuracy. He is about 7 feet high, has broad hands and feet and his body is covered by a prolific growth of hair. In short, he looks like the very devil.

The above article is more of a patchwork of reporting than merely relying on one account. As such it is far more accurate than the McCulloch and Page sighting of 4 years previous. The 9 foot height was clearly an exaggeration (though near 7 feet is still huge). "Something after the fashion of a gorilla" is far from anything ape-like other than a hairy body and long arms. In any case, the miners knew what a gorilla looked like. The fact this "animal-man" walked upright with apparent ease, which no gorilla can do, was what was so surprising to those who saw it. No one was and is expecting to see a human form and yet a beast.

Unbeknownst to the good folks of Oregon, in 1846 Alexander Caulfield Anderson encountered an entire party of 'wild men' entrenched in the mountains of British Columbia that behaved in the same way. He wrote in his journal that while his expedition was attempting to find a new pass through the mountain over Harrison Lake they were attacked by these 'hairy giants,' who hurled stones

---

[5] It is interesting to note that in this one paragraph it is called "monster," "animal" and "man."

at them until they were forced to retreat. Such organized behavior clearly seemed to be human but primitive. And, indeed, the Indians reinforced this upon Anderson by telling him of "hairy Indians" who lived in the mountains.

So human was another "wild man" that it was, in fact, thought to be a long lost hunter, John Mackentire of Lebanon, Oregon. This location is about equidistant between Mount St. Helens and the Sixes, and one wonders therefore if this "wild man" was not the 'Kangaroo Man' himself. The *Nevada Morning Appeal* for 31 December 1885 reported the details. Hunters came suddenly across this "wild man" while he was eating a killed deer raw. Taken by surprise, 'he' then darted off with inhuman agility and disappeared into the forest. The encounter caused a lot of "excitement" because the "wild man" was thought to resemble Mackentire, who had mysteriously disappeared on a hunting trip four years before in 1881 when he separated for a short time from his partner. Despite an extensive search he was never found. However, because this wild man was so human they thought perhaps it was Mackentire having gone completely demented after having been lost all those years. By a somewhat stretch of illogic they thought he could have grown a body coat of hair in the wilds.

Obviously, none of the men got a good look at the "wild man." But they saw something they could identify with a hairy human. Both this and the unusual speed of the "wild man" once again eliminate the notion of a short-legged, waddling ape.

The Clallam had a similar regard for an enclave of "wild men" in the Cascades as the Salish had for the Saskahaua Georges in British Columbia. They called them the Seeahtik tribe. They said they were offshoots of the Klickitats, a tribe who were, naturally, their enemies. In any case, Seeahtiks were like Saskahaua Georges inasmuch as both tribes were manlike, hairy, smelled foul, hurled rocks, and attacked cabins.

The aggressive instincts of the "wild men" and their reputation for "cannibalism" add chilling dimensions to an encounter in Idaho. It occurred during the same period of time in which the Sixes River mining operations was being vexed by the 'Kangaroo Man.' *The Chronicle* of Wilkesboro, North Carolina, along with other papers, picked up on the story and ran the entry on 5 February 1902 as an oddity:

The residence of the little town of Chesterfield, located in an isolated portion of Bannock County, Idaho, are greatly excited over the appearance in that vicinity of an eight-foot, hair-covered human monster. He was first seen on January 14, when he appeared among a party of young people, who were skating on the river near John Gooch's ranch. The creature showed fight, and, flourishing a large club and uttering a series of yells, started to attack the skaters, who managed to reach their wagons and get away in safety. Measurements of the tracks show the creature's feet to be 22 inches long and 7 inches broad, with the imprint of only four toes. Stockmen report having seen his tracks along the range west of the ringer. People in the neighborhood, feeling unsafe while the creature is at large, have sent 20 men on its track to affect its capture.

There is no other way to interpret the action of the "human monster" except as predatory, an undeniable element both in Bauman's story to Roosevelt (occurring also in Idaho) and in the story of the Skoocooms. Despite its "human" appearance, its 4 toes make it plain it was not human in any way. Because of this and its size, the Bannock County sighting stands as an exception. But it does not stand aloof. It helps to interpret a chilling encounter in Washington State.

This incident happened in 1924, when it seems there was a genuine attack by the legendary Skoocooms on Mount St. Helens. Over the night of July 11-12 five miners were besieged in their cabin, which overlooked a canyon near Kelso, Washington, on the shoulders of the mountain of the "race of beings of a different species." Despite the miners repeatedly firing their guns, the attackers would not flee but continued their assault, pounding the cabin with boulders, hurling smaller rocks, and trying to break in the door and roof. The next morning revealed that the ground was imprinted with hundreds of footprints up to 19 inches long and with only 4 toes. The canyon, which had no name before this, would be named Ape Canyon afterward, a name which it still holds to this day based not on the descriptions the men gave of their strange, hairy assailants but on the local vague legend that "gorillas" already existed on the mountain.

Before noon the five men had made it back down the mountain, stopping first at the Spirit Lake Ranger station. The ranger on duty, Bill Welch, was greeted at his barn by an excited miner with a rifle slung over his shoulder. Thirty-nine years later, in the *Longview Times*, Welch recalled the bizarre moment. "The miner was pretty wild-eyed, and I was worried how this fellow was going to behave." The miner stopped before him and declared: "Well, I got 'im."

"Got who or what?" replied Welch.

"The mountain devil," the miner replied.

"You mean a cougar?"

"No, the mountain devil," replied the nervous man.

"You mean a wolverine?"

"No, the mountain devil."

Welch was leery. He remembered the miner from a few weeks before when he came around the station asking about a fire permit. It was then that the miner had also told him the "mountain devils" had been trailing and bothering them for years. He hadn't seen them but he had seen their tracks. Welch had actually invited this visit by having assumed (without expressing it) that the mountain devils were wolverines. He had told the miner that if he should "run into them again" he was to let Welch know.

On this day the miner was doing just that. Now he told Welch his partners were in the car by the station. They were going home and "they would never come back." Now Welch really thought this guy had "blown his top." But when he walked over and saw the other men sitting in the car: "they were just as wild as he was, clutching their guns."

When they reached town and told what happened the local news stringer picked up on it and phoned Portland the byline. The Portland *Oregonian* ran the story under the header FIGHT WITH BIG APES REPORTED BY MINERS. Here's the story.

The strangest story to come from the Cascade Mountains was brought into Kelso today by Marion Smith, his son Roy Smith, Fred Beck, Gabe Lefever and John Paterson, who encountered the fabled 'mountain devils' or mountain gorillas of Mount St. Helens this week, shooting one of them and being attacked throughout the night by rock bombardments of the beasts.

The men had been prospecting a claim on the Muddy, a branch of the Lewis River about 8 miles from Spirit Lake, 46 miles from Castle Rock. They declared that they saw four of the huge animals, which were about 400 lbs. and walked erect. Smith and his companions declared that they had seen the tracks of the animals several times in the last six years and Indians have told of the 'mountain devils' for 60 years, but none of the animals ever has been seen before.[6]

Smith met with one of the animals and fired at it with a revolver, he said. Thursday Fred Beck, it is said, shot one, the body falling over a precipice. That

[6] 80 years if we count Kane's diary entry for 1847.

night the animals bombarded the cabin where the men were stopping with showers of rocks, many of them large ones, knocking chunks out of the log cabin, according to the prospectors. Many of the rocks fell through a hole in the roof and two of the rocks struck Beck, one of them rendering him unconscious for nearly two hours.

The animals were said to have the appearance of huge gorillas. They are covered with long, black hair. Their ears are about 4 inches long and stick straight up. They have four toes, short and stubby. The tracks are 13 to 14 inches long. These tracks have been seen by forest rangers and prospectors for years.

The prospectors built a new cabin this year and it is believed it is close to a cave thought to be occupied by the animals. Mr. Smith believes he knows the location of the cave.

Elaborating on their experience years later, Fred Beck explained how they were stalked by the creatures for a couple of years before the incident. On one occasion Beck, in company of Marion Smith, saw one behind a tree watching them. They finally had enough. Smith fired at it. He was sure he shot it 5 times, but when they tracked it they never came upon a body. For most of the time their mining operations had been ongoing the miners had also heard distant "whistling."

This came to a crescendo the night of the attack. They again heard the crunching of footsteps in the underwood as they walked back to the cabin. Then while inside they continued to hear the whistling noises until dark; but they were louder and closer than usual. Then the cabin was suddenly attacked. Beck recalled it began with a hairy arm breaking through the chinking of soft pine that was stuffed between the log beam walls of the cabin. This fell on Marion's chest while he lay in his bunk. He jumped up and out of the way as the hand of the hairy creature grabbed a nearby ax by the handle and started to withdraw it out the crevice. Smith quickly grabbed it and turned it so the hairy arm could not pull it through. He then grabbed his rifle and shot the handle. The hairy hand immediately let go. After this, large boulders were thrown down on top of the roof from the cliff overhead while other creatures shook the cabin and tried to get in. Then they jumped up on the roof and started running around.

Although Beck was recalling this years later in a taped interview, he still excitedly remembered the events of that night. "Well, I want

*No mountain has attained such lore in Bigfoot history as Mount St. Helens. Its taboo Indian history as a dwelling place for "a different species of men" is mixed with vague White Man legends of "mountain gorillas" and frightening encounters with hairy creatures which can be neither men nor apes.*

to tell ya, pretty near all night they were on that house, trying to get in, you know. We kept shootin'. Get up on the house we'd shoot up through the ceiling at them. Couldn't see them up there, you could hear them up there. My God, they made a noise! Sounded like a bunch of horses were running around up there."

In an interview in 1969 Beck would later describe them. "Well, they was tall. . .they looked to me like they was eight-foot tall, maybe taller, and they was built like a man, little in the waist, and big shoulders on, and chest, and their necks was kinda what they called bull necks. . ." He said they never went down on all fours, but remained bipedal. They had long arms and were hairy all over. He estimated they were about 600 to 800 pounds, considerably bigger than what they told the newspaper in 1924.

As they were going up to the mine the next morning to get their tools in preparation to pull out, one crossed the road. That's when Beck shot it three times in the back and it fell over a precipice and went down into the canyon. They never did know if he killed it or not.

Beck admitted that the newspaper reporter never believed them. After they hit town the reporter bugged them for the story, especially after hearing that "detectives" were coming from Portland.

When the "detectives" did arrive, they went straight up to the cabin (the reporter did not) and measured the tracks. It had rained the night before, but there were fresh tracks yet again made after it rained. They measured these at 19 inches.

There is obviously much confusion in the retelling. Equally obviously, there was much talk by the miners before the newspaper caught wind of it. The notion that "detectives" came from Portland must be based on Bill Welch's investigation. Welch knew where the miners had built their cabin, so he pursued this further with district ranger Jim Huffman. Huffman, too, thought they must have been referring to wolverines, a damnable and ferocious nuisance that destroys anything it doesn't eat. After the article came out describing "miners fight with big apes" two journalists also arrived, Seattle newspaperman Frank Lynch and freelancer Burt Hammerstrom (Clarence Darrow's brother-in-law). They went with Welch and Huffman to investigate and instead of finding wolverine tracks all they could do was verify the miners' reports. Huge footprints surrounded the tattered and beaten cabin. It was a strange footprint, as much as 19 inches long, each showing 4 toes. Stones littered the area, but it seems that most of the commotion and damage had been done by the "creatures" dancing around on the roof.

In retrospect it seems likely in the excitement that the prospectors were simply missing their "assailants" with their gunfire, or perhaps when they got to town they engaged in some harmless boasting. There is also that blurring force of time. Memory contamination happens to us all, and by 1965 and then 1969 when Beck was again interviewed over 40 years had elapsed.

The Ape Canyon Incident has gotten wide circulation, and it is unquestionably authentic in its basic elements, though today a few dimensions and elements of the creatures might be exaggerated and chronology confused. But it did happen, and it should be considered more than just the material of campfire stories. Basic elements of it are verified more by other incidents rather than by taking the story at face value. The Bannock Co. sighting was obscure by 1924 and so was the incident reported by Caulfield Anderson in 1846 and Kane (on the very mount) in 1847. All of their accounts reported or implied manlike creatures capable of brandishing clubs or hurling rocks, and for Bannock County's report the assailant also had 4 toes.

That creatures of that size can band together like a wolf pack and not only attack but engage in revenge (for Smith shooting at them earlier) should strike one as truly frightening. It is certain that if the men did not have the refuge of the log cabin they would have been massacred. They do indeed seem to have encountered the hated Skoocoom, "the race of beings of a different species, who are cannibals."

Mount St. Helens has become legendary in the world of Pacific Northwest folklore. The Ape Canyon Incident was something that gave form to years of vague legends. But what are the Skoocooms? Their footprints preclude them being apes; but they also preclude them from being humans. Only with White Man did the "mountain devils" begin to be commonly called "mountain gorillas" and then after 1924 "hairy apes." But the one tangible proof— their strange big footprints— say they are neither.

The same enigma is found in a later sighting. On August 19, 1933, a strange "shaggy appearing human" was once again seen in Oregon. This sent a "veteran party of woodsmen on its trail." However, it left no trace and was written off as a "lunatic, hermit, or figment of the imagination." This case, however, has an interesting epilogue. Three vacationers were then lost in the area, requiring a rescue party to go looking for them. Mr. Irving C. Allen, a city attorney of Seaside, Oregon, became a witness. "I happened to lag behind the searching party on Bald Mountain. I glanced behind me, and there it was. It looked like a shaggy beast, yet human. It was growing dark, and I couldn't see it plainly. As I stood watching, it turned and fled into the darkness."

A few days later Clement Klink was with another search party. He reported: "We were hunting near the 'death trap.' I looked up the cliff and saw an animal-like human peering down at me. I watched the creature, which was at a considerable distance, and it bounded away."

Perhaps to add skeptical endorsement to the whole idea, the newspaper concluded with the pithy comment of William Laighton, "who knows every animal trail and crag on Tillamook Head, and reported to have seen the wild man. 'Well, I saw something queer,' Laighton said when questioned. He refused to say more."

The newspaper does not mention whether the three lost vacationers were ever found.

The most frustrating thing once again about this sighting is that of its description of an "animal-man," something that suggests a "creature" far less than gigantic. By 1933 surely every one of those men quoted above knew what a gorilla or ape looked like. If they had clearly seen an ape, why would they not say so or express that perhaps a gorilla escaped from a carnival? Instead they are not able to advance more than what the three original witnesses said: "he was a 'shaggy-appearing human,' with an animal-like face, who bounded away when observed"— identical to the Sixes River Kangaroo Man that had vexed the area more than three decades before. The "wild men" of 1885 and 1933 were different from the Skoocoom only in size (apparently) and having 5 toes— in other words, far more human; and yet the Skoocoom behaved in a human way inasmuch as they attacked the miners in a concerted effort to kill each one. We come back again to the notion of "two tribes."

Encounters with strange hairy creatures in Alaska and the coast-line of British Columbia add injury to insult when trying to determine what the "animal human" is. Here the creature was referred to as the "Bushman."

In an area east of Thomas Bay, Alaska, in the year 1900, a prospector was attacked by three hair-covered creatures. He ran for his canoe and was able to push off before they could catch him. He later described them as "the most hideous creatures. I couldn't call them anything but devils, as they were neither men nor monkeys, yet they looked like both. They were entirely sexless, their bodies covered with long coarse hair, except were scabs and running sores replaced it."

Although the above story is vague, and therefore one can question its source, there is no doubt about the story of Albert Petka in 1920. He lived on his boat and was attacked by 'the Bushman,' which his dogs eventually drove off. However, Petka later died of his injuries. In a similar vein in 1943 at DeWilde Camp, near Ruby, Alaska, John Meyer— the "Dutchman"— was able to reach a near-by settlement in his boat where he reported that he had been attacked by the 'Bushman' but that his dogs had chased it away. Although Meyer was in pain, neither he nor the settlers realized how hurt he was. Shortly afterward Meyer died. It was discovered to have been from internal bleeding.

It is from these early reports that a believable image of a real creature(s) can be found. These reports of encounters with "animal humans" were isolated and not common campfire gossip. They became only indistinct legends and remained so because no one would follow up on the reports with a serious attitude. Few reports ever made it as far as the newspapers. But those that did were long before any hype. These articles, though contradictory and containing some reportorial gin, serve as a sure foundation. Long before Burns' 1929 article they reinforce a human size and aggressive nature to "something." Collectively they also show why this creature inspires fear like nothing else can, why it is hard to describe except as "hideous," why it is a "wild man" and yet not Man, and they confirm the most bizarre feature: that 'it' walks on its hind legs like a man.

It is this quality which perhaps is responsible for why there is no consistent description of its face to be found— all witnesses were surprised by the general appearance of a manbeast. Yet it was not just beast. It displayed the instinct of a degenerate man and as such it was not just an animal to be shot but a cunning and unpredictable predator to fear. It was a terrifying creature that could not really be identified because it was so rare, so frightening, and encounters with it so unexpected, that no witness was ever capable, understandably, of taking down details. Not even the gorilla has maintained such an enigma.

Although the 1933 article appeared in a number of newspapers as an oddity or filler, such as in the above account in the *Visalia Times Delta*, the idea of some kind of hairy "animal human" never became widespread, even in the Pacific Northwest. One can imagine this is because the "thing" was still so nebulously described it had no real "handle" by which to promote the idea. The Abominable Snowman, on the other hand, is quite a moniker. The explosion of this concept upon the world, though it was centered a world away from the Pacific Northwest, was actually the crucial stimulus to research the existence of the "what is it?" in America.

# Chapter 2

# Fourteen Thousand Feet

to

# Earth

Curiously, around the same time as frontier Americans and Canadians were hearing vague stories of "animal humans," the British in Nepal were hearing about something equally nebulous—the "wild man of the snows." Similarly, the local Nepalese Sherpas called it Mih-Teh, which roughly translated would be "man thing." And, similarly again, it was said to walk on two legs, to be hairy and often violent. It dwelt in the rocky heights of the Himalayas, around 14,000 feet up, above the tree line and below the perpetual snowline, and it would climb higher or descend lower based on the summer and winter weather changes.

Coincidence, as we know, is one of the most powerful influencing factors in any investigation. Uncovering similar stories in completely removed and unconnected peoples temps even the purest humbug to consider that there is a kernel of truth underlying all the accounts. And indeed at first glance the "wild man of the snows" sounds like it could be the same "wild man of the woods" the Indians in America had reported. Theories abound about ancient migrations across the Bering Land Bridge to explain many Old World species that are found in America. Thus speculation along these lines is not out of place. However, with a little study it becomes obvious that it is in name only that there is coincidence. Instead of the combined Eurasian and American stories together revealing the single entity behind it all, they reveal nothing but contradictions.

For instance, the "wild man of the snows" is so inconsistently described several solutions have been offered as its reality. The first Europeans tended to regard 'it' as various animals. In the very first mention of 'it' in western literature (*On the Mammilia of Nepal,* 1832), B. H. Hodgson, British Resident of the Court of Nepal, noted that his coolies reported back to him about being attacked by a hairy *rakshas* (demon). From their descriptions, and for lack of a better explanation, Hodgson thought they had seen an orangutan. In 1889 a Major L.A. Wadell heard of something called the "Yeti" when he came across large footprints at the 17,000 foot level in the northeastern Sikkim Himalayas. Wadell was so piqued by the sight of the tracks and by the accounts of the Sherpas that he is perhaps the first to actively try to investigate it. In *Among the Himalayas* he wrote that "The belief in these creatures is universal among Tibetans, none however of the many Tibetans I have interrogated on the subject could ever give me an authentic case. On the most superficial investigation it always resolved into something that somebody had heard tell of." As a result, he ended up believing that "these so-called hairy wild men" must have been the great yellow snow bears.

When the world first took notice in 1921, an entirely new entity emerged. The great mountain climbers had arrived in Nepal to try and tackle Everest, and the leader of the 1st Everest Reconnaissance Expedition, Lt. Colonel Charles K. Howard-Bury[7] encountered something surprising between 20 to 21,000 feet altitude on the Lhakpa-La Pass. He records with an air of surprise:

Even at these heights we came across tracks in the snow. We were able to pick up tracks of hares and foxes, but one that first looked like a human foot puzzled us considerably. Our coolies at once jumped to the conclusion that this must be 'The Wild Man of the Snows,' which they gave the name as Metoh kangmi, the Abominable Snowman who interested the newspapers so much.

This great interest from the newspapers stemmed from a telegram Howard-Bury had sent to Darjeeling after he came down from the mountain. In his report metoh was mis-transcribed as metch. Kang (snow) Mi (man) was easy to translate. Metch, however, was ambiguous to everyone. Henry Newman, an Englishman

---

[7] His account was published in *Mount Everest Timeline,* and it is on page 141.

familiar with the dialects of Nepal, was called upon to translate. He said it meant abominable. Therefore Howard-Bury was referring to an "Abominable Snowman." Newman was more captivated by the implications than others. As soon as some of the porters arrived in Darjeeling he went and questioned them. Stories about the metoh kangmi bordered on fantasy. They had backward turning feet and long hair falling over their eyes. They kidnapped women and children and were genuine mountain ogres. "The whole story seemed such a joyous creation," he wrote in a letter to the *Times*, "that I sent it to one or two newspapers. Later I was told by a Tibetan expert that I had not quite got the force of the word 'metch' which did not mean 'abominable' quite so much as filthy or disgusting, somebody wearing filthy tattered clothing."

It was too late. The British Press had jumped on the moniker. The whole idea of a ghoulish "abominable" man roaming the mountain snows giving everybody a right scare delighted the public. It also gave them a "wild man," not a beast. Newman quickly speculated along these same lines. "This, I am convinced, is that in Tibet there is no capital punishment, and that men guilty of grave crimes are simply turned out of their villages or monastery. They live in caves like wild animals, and in order to obtain food become expert thieves and robbers. Also in parts of Tibet and the Himalaya many caves are inhabited by ascetics and others striving to obtain magical powers by cutting themselves off from mankind and refusing to wash."

Charles Howard-Bury, who innocently had started it all, remained skeptical about all the excuses and later believed he might have seen the footprints of a large loping gray wolf. This he thought was possible if its hind feet had stepped into the tracks of its forefeet imprints. Howard-Bury convinced himself of this after reading Newman's tales about the Kangmi ogre, all of which definitely smacked of folklore. Obviously, the actual evidence for the "Snowman" was so nebulous that a gray wolf, a bear, an orangutan or a raggedy Andy mystic on-the-outs could be viewed as its reality.

A gray wolf certainly didn't carry the zing to keep a new legend alive. The large human footprints were a better bill. The Abominable Snowman thus became a rugged chap, some kind of large caveman we might expect to be happily residing in a Rice-Burroughs novelette. The form in which it subsequently spread was indisputably that of a British music comedy hall or tabloid man-beast.

This image only tarnished any believability thereafter. By 1938

the time had come to take on the whole legend. The scientific proffering of F.S. Smythe's highly publicized debunk almost totally
eroded public confidence in the legend. In the summer of 1937 he
had come across the "imprints of a huge foot, apparently of a biped" at the 16,500 foot level. His porters declared it to be that of a
mirka —the Snowman— but Smythe was sure it was that of a bear.
His arguments won the day. A poll in the fall of the year in Britain
awarded the prize to bears as the reality behind the Abominable
Snowman.

This was too ignoble a fate for a sensation that had enchanted
the world since 1921. With good British sportsmanship H.W. Tilman in his *Everest—1938* [8] hit home the point that "it will be seen
after reviewing the evidence which I shall marshal for and against
the existence of the 'Snowman' that except for one instance of small
value everything turns upon the interpretation of footprints."
Obliging the reader, Tilman then provides us some interesting examples on how difficult it was to interpret tracks in snow. He reports the incident where Ronald Kaulback and his porters came
across 5 sets of tracks in the Upper Salween above the 16,000 foot
level. They seemed identical to a bare-footed man's. His porters
disagreed. Two said they were a snow leopard's, "two thought they
were those of 'mountain men' which they described as like a man,
white-skinned, with long hair on head, arms and shoulders."

Ever the gentleman, Tilman had nonetheless made his point that
nobody's interpretation of footprints is the final word. In his post
scriptum he let the legend stand. "And I think he would be a bold
and impious skeptic who after balancing the evidence does not de-
cide to give him the benefit of the doubt." Until some known animal can be proven to be the Snowman, why not have a genuine
abominable chappie around the mountains of the Himalayas?

*Everest—1938* had come at an interesting time. It was published
in 1948, delayed 10 years because of the war. The Abominable
Snowman had almost disappeared in those beleaguered years. The
last image Britons had was the poll of 1937 writing it off as a bear.
Surprisingly, though only an appendix was devoted to the Abominable Snowman, Tilman's book revitalized the interest of a post
war Britain tired of calamitous news.

Credit for this must really go to Tilman's epilogue. Just before

---

[8] His account of his unsuccessful 4th Everest Expedition. He devoted only an appendix "Anthropology or Zoology?" to the Abominable Snowman.

publishing he was made aware of an old (1925) but dynamic report by N.A. Tombazi, a British–Italian photographer and a member of the Royal Geographic Society. He was able to squeeze in some of the report, but it is best to quote Tombazi's actual report verbatim.[9] Tombazi was at the region of the Zemu Glacier, at 15,000 feet altitude, when his Sherpas reported something in the distance. He put his field glasses to his eyes.

The intense glare and brightness of the snow prevented me from seeing anything for the first few seconds; but I soon spotted the 'object' referred to, about two to three hundred yards away down the valley to the east of our camp. Unquestionably, the figure in outline was exactly like a human being, walking upright and stopping occasionally to uproot or pull at some dwarf rhododendron bushes. It showed up dark against the snow and, as far as I could make out, wore no clothes. Within the next minute or so it had moved into some thick scrub and was lost to view.

Such a fleeting glimpse, unfortunately, did not allow me to set the telephotocamera, or even to fix the object carefully within the binoculars; but a couple of hours later, during the descent, I purposely made a detour so as to pass the place where the 'man' or 'beast' had been seen. I examined the footprints which were clearly visible on the surface of the snow. They were similar in shape to those of a man, but only six to seven inches long by four inches wide at the broadest part of the foot. The marks of five distinct toes and instep were perfectly clear, but the trace of the heel was indistinct, the little that could be seen of it appeared narrow down to a point. I counted 15 such footprints at regular intervals ranging from 1 ½ to 2 feet. The prints were undoubtedly of a biped, the order of the spoor having no characteristics whatever of any imaginable quadruped.

Dense rhododendron scrub prevented any further investigations as to the direction of the footprints, and threatening weather compelled me to resume the march. From inquiries I made a few days later at Yoksun, on my return journey, I gathered that no man had gone in the direction of Jongri since the beginning of the year. The coolies naturally trotted out fantastic legends of Demons, Snowmen. Without in the least believing those delicious fairy-tales, notwithstanding the plausible yarns told by natives, I am at a loss to express any definite opinion. I can only reiterate with a sufficient degree of certainly that the silhouette of the mysterious being was identical with the outline of a human being.

---

[9] 1925, published by Maxwell Press of Bombay— *Account of a Photographic Expedition to the Southern Glaciers of Kachenjunga in the Sikkim Himalaya*.

It is, of course, impossible for a bear to bend over and grab and pull at anything. Only a primate has hands with which to clutch and grab. In fact, so human was the image that Tilman ended his book telling us Tombazi leaned toward Newman's explanation of the "filthy mystic."

Therefore by 1948 an interesting and confused collage had been established independent of footprints. The gamut ran from a hairy ferocious animal like a bear, mountain men that sounded straight out of a *Land that Time Forgot*, and something very human but animal.

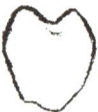

Only 3 years later in November 1951 a defining encounter would change it all. Expert mountain climbers Eric Shipton and Dr. Michael Ward were on the 1951 Everest Reconnaissance Expedition scouting a new way up for the upcoming Everest Expedition. While on the Menlung Glacier with their guide Sen Tensing they too came across and then followed peculiar, large tracks. Seldom ever deferred to on this extraordinary find, Shipton nevertheless explained it all in clear detail in *Menlung La*, confirming circumstances of which most are unaware.

Sen Tensing was the only one of the three of us who had no doubts. He immediately pronounced the tracks to be those of the *Yeti* or "Abominable Snowman." He told us that two years before, near Thyangbochi, he had seen one of these creatures at a distance of some 25 yards; he described it as being the height of an average man, tailless, with a tall, pointed head and covered with reddish-brown hair except on its face, which was bare. I had known Sen Tensing intimately for a very long time; I had a considerable opinion of his intelligence and his sincerity and I was not inclined to doubt his story. As for his explanation of these tracks, though I did not dispute it openly, for the time being I reserved my judgment. . .

The snow where Shipton first met the tracks was deep; consequently, the tracks were "nothing but shapeless holes." However, due to his respect for Sen Tensing's opinion, he decided they would divert and follow them.

. . .As we went down the glacier, however, the snow became less deep and the footprints more regularly shaped. At length we came to places, particularly near crevasses, where the snow covering the glacier ice was less than an inch thick. Here we found specimens of the footprints so sharply defined that they could hardly have been clearer had they been carefully made in wax. We could tell both by comparing one print against another and by the sharpness of the outline there had been no distortion by melting, and from this we inferred that the creatures (for there had evidently been two) had passed only a very few hours before. They were going down the glacier. The footprints were 12 inches long and about 6 inches wide. There was a big, rounded toe, projecting very much to the side; the middle toe was well separated from this, while three small toes were grouped close together. There were several places where the creatures had jumped over small crevasses and where we could see clearly that they had dug their toes into the snow on the other side to prevent their feet from slipping back.

For scale Shipton placed the end of an ice pick next to the print in one photo and in the other Michael Ward placed his booted foot next to it. The photographs were perfect and clear from a camera almost dead-center over the print. The Shipton photograph of the clearest print became instantly famous, so much so that when Shipton arrived at Karachi Airport he was overwhelmed by reporters. This photo today is still the most important proof that the "Yeti" does exist.

*Exhibit A of evidence: The Shipton Print, showing the left foot from the bottom.*

But what exactly is it? It isn't what Tombazi reported. It certainly isn't human. And definitely no bear. Despite every effort to try and provide an alternative explanation for the footprint (that is, other than Yeti), what Eric Shipton photographed has never been explained. The Press soon dubbed it the footprint of the legendary "Abominable Snowman," exactly what many people still believe it to be. But what is it? The Abominable Snowman finally materialized as something no one had expected. . . .And it wasn't long before someone dropped the bombshell and said "apeman."

This controversial enigma dogged the most publicized expedition in history— the 1954 *Daily Mail* sponsored Everest Yeti Expedition— to the chagrin of its two leaders, Ralph Izzard and Charles Stonor. Yet despite Stonor and Izzard's attempts to distance their quest from the world of Rice Burroughs, it seems certain that this major British newspaper never would have backed such an expedition if this sensational discovery did not possibly lay behind it. The *Daily Mail* had commissioned Izzard, their local foreign correspondent, and a combined group of zoologists and hundreds of Sherpas to track up the High Himalayas for 6 months in hopes of nothing less than the story of the century. Syndicating Izzard's weekly reports put the paper back on top, as readers in 24 countries eagerly waited for the next edition's evolving details.[10]

Izzard's captivating book *The Abominable Snowman Adventure* was complimented by *The Sherpa and the Snowman* by Charles Stonor, the expedition's biologist, both published in 1955. Together Izzard and Stonor establish for the first time a believable image of the Yeti, stressing always that it is but an animal. From Sherpa stories both had become convinced that "indications seem stronger than ever," as Izzard would write, "that we are dealing with 'Animal X,' an unknown variety, or possibly a species, a dangerous beast of marked ferocity, who has little to commend him other than his rather endearing name."

In his preliminary report, Stonor declares: "It is said to be about the size of a fourteen-year-old boy, of the same build as a man; covered with light reddish hair, a little lighter on the chest, and with the hair longest about the head and the waist. The head is strikingly pointed. It has a loud wailing yelping call when it is heard near at hand, often makes a chattering noise. Its call is frequently heard at night."

Stonor also clearly established for us the reason for its name. Yeti (pronounced Yay-tay) merely means teh (entity, thing) of the yay (rocks). As the name implies, the Yeti lives entirely "in the huge region of rocky country which is too high for trees to grow and below the perpetual snowline." Although many animals dwell in this area (and are thus called Yeti), Stonor clarified: "The *Mi-Teh* is the Teh *par excellence* of the Sherpa country and, when speaking of the *Yeh-Teh*, they normally mean this beast: so that the names are

---

[10] The team consisted of Ralph Izzard, biologist and anthropologist Charles Stonor, zoologists Dr. Biswamoy Biswas, curator of the Mammals Section of the Calcutta Museum, Gerald Russell, Tom Stobart, expert mountain climber John Angelo Jackson, and medical doctor Bill Edgar.

for all practical purposes interchangeable."

Throughout his research, however, Stonor had to come to grips with this puzzling word: "The people do not seem clear as to why it is called the Mih (man)—Teh: by no stretch of the imagination do they look upon it as a man— it is *an animal and nothing else.*"

Stonor quickly accumulated information on incidents that illustrated his point "all at first-hand and given to me by the individuals concerned in the incidents." A 1947 incident involved a man named Dakhu of Pongboche village. He was up in the mountains searching for one of his yaks when he lost his way in the rocky area. When trying to find his way back he saw a hairy animal among the rocks and called out to it thinking it was his yak. He called out three times, at which point the thing stood up. He realized it was a *Yeh-Teh*. With calculated movement the thing advanced toward him a few paces, walking on two legs like a man "and started pulling up tufts of grass with its hands as if agitated." Dakhu was beside himself and went tumbling down the hill to escape. He never saw one again, but claimed he has heard their mewing, barking calls at night.

Another incident happened in 1950 to Lakhpa Tensing when he went up the slopes from his village in order to round up some of his stray yaks. He got lost among the rocks and was trying to find a trail down when he heard something like a puppy yelping. He homed-in on the sound and finally came upon the "mangled corpse of a freshly killed pica and looking further on he saw, only thirty yards distant, a *Yeh-Teh* sitting on a rock." He saw it only from behind, and estimated it was the size of a boy. It had reddish colored hair and a "very pointed" head. He carefully and quietly crept away. Despite how many times he was asked, he would not elaborate. That is all he saw.

The pointed head is the most declared characteristic. But the most unusual feature of its cone head is a crest of bristly hair.

A very clear description of this comes from a Sherpa named Mingmah. In 1949 he was out tending his yaks on the mountain slopes when he heard a call. Assuming it was his friend tending his yaks on the plateau he shouted back "None of your yaks have come this way." The yelping and howling continued and then he saw something walking down the side of the rocky slope coming for him. It was a Yeti! He was terrified and locked himself into a herdsman's hut.

Outside the creature continued to prowl and howl about. Mingmah looked out a broad chink in the wall and there saw the Yeti only a few paces away. "It was about the size of a teenage boy, very thick-set and the general shape of a human being. It was covered with rather short and reddish-brown hair, which seemed to be stiff, as the hair on the upper part of the body all pointed upwards, while on the lower part it sloped downwards. The color was paler on the lower part of the chest. The feet looked especially hairy. The hands looked like those of a man. The head was very pointed and the hair on each side sloped back as with a man. There was a narrow strip of short very stiff-looking hair running up the forehead and over the top of the head, apparently like a crest. The hair on the head was not long. The face was bare, flatter than a man, but not so flat as a langur monkey, and was a dark brown color."

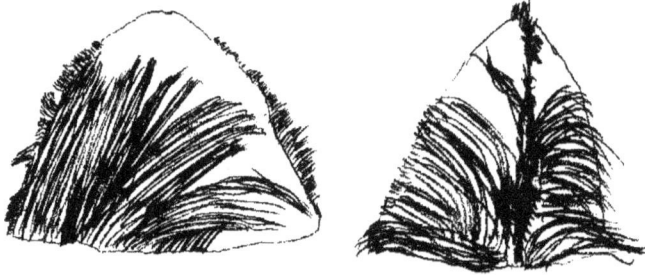

*Two monasteries hold reproduction Yeti scalps. The Pangboche Scalp on the left and the Khumjang Scalp on the right. One is possibly 300 years old. Both represent the Yeti and both contain the bristly crest. In 1960, Edmund Hillary made a big splash about debunking them as genuine, but Stonor had already clarified that they were reproductions made by the monks. A member of an Indian expedition in 1952, a Mr. Parikh, was allowed to remove some hairs from the Pangboche scalp. Zoologist Dr. Leon August Hausman, one of the leading authorities on hairs of any kind, examined them at his facility in New Jersey. Under the microscope the hair shafts' structural elements agreed with those of a bear or anthropoid. However, they matched no known species in Tibet. At the very least, these famous caps impressed upon us that Yeti had a tall cone head.*

When the Yeti saw him looking at him, it bared its teeth and snarled. Mingmah recoiled at the size of the teeth. "It moved about on two legs, taking long strides, and rather stooping, with its arms hanging down by its side." Mingmah grabbed a smoldering firebrand from the hearth and jabbed it outside the chink hole. Only then did the Yeti make off.

Stonor set out to establish its diet as one way to anticipate its movements. He had repeatedly heard it consisted of small rodents among the rocks. "There is widespread belief," he wrote, "that it catches its prey, bangs it on the rock to kill it and then disembowels it, leaving the innards, and eating the rest. Two yak-herds have told me how they found the very fresh innards of a marmot among the rocks and could see two large footprints nearby." While in the country west of Namche Bazar he, too, came upon the "fresh innards of a marmot." There was a little fur nearby and nothing else. He writes: "*Something* must have killed it. Tibetan foxes are common and feed on such animals: but I have yet to hear of a fox disemboweling its victim and thereafter carrying it off, leaving its guts behind."

The Sherpas confirmed that the Yeti also sometimes swallows clay (which Stonor thought was for bulk). It might also snatch young yaks, musk deer and young tahr, plus bird's eggs. On two occasions Stonor came across feces that contained the remains of small rodents, fur, and a small amount of earth. From the local fauna only two things could have been responsible: the snow leopard or a *Yeh-Teh*. The Sherpas said that a leopard's feces would include deer and tahr. They were sure this was from the Yeti.

The purpose of these vignettes, of course, was to hit home the point that Yeti was but that Animal X. Underscoring this, both authors made it clear that the Yeti is also reported to go down on all fours when in a hurry or when chasing cattle.

Stonor and Izzard were clearly out to make a point. By the time the expedition was in the planning stage the "subhuman" theory had already lent that old music comedy hall beast image to their quest; and it was from this notion that he and Izzard wished to distance themselves vociferously. But however passionate they were, Izzard and Stonor did not persuade. Everything about the encounters spoke of Pithecus X, an unknown type of primate. That being established, it opened the door to evolutionary debate.

Visually, the Shipton footprint had struck almost everybody as an ape's foot evolving into that of a human foot. Even one of the expedition's zoologist, Gerald Russell (who had assisted in the capture of the first live panda), was very pro the idea that the Yeti could be a subhuman. Joining such ranks were several other respected zoologists and anthropologists in Britain. For instance, J.P. Mills, past president of the Royal Anthropological Institute and an advisor to the British government on the Himalayas, went public about his ideas. In written support of the expedition he noted its potential to advance the knowledge of mankind. "The evidence in support of a Yeti is very strong indeed and seems to indicate that it is either an ape more man-like than hitherto known or even a very primitive species of man. . .Whether the Yeti is a remnant of such a race, preserved by the absence of natural enemies and the inaccessibility of its habitat, remains to be proved. Should the expedition bring back such proof they will have made a contribution to the study of man and his ancestry beyond the power of words to describe."

As the *Daily Mail* expedition's fame grew, the significance of the Yeti in theorizing grew with it.[11] Izzard tells how British naturalist Earl Nelson laboriously clarified in the *Evening News* that the Yeti will not prove to be a man, sub-man or even ape-man. Rather, the Yeti could prove to be a lateral descent from a common simian ancestor. Nelson declared: "It is now clear from the fossil record that at various time periods in the remote past a great many different species of men were evolved, just as happened in the case of other mammals, notably the horse and elephant." While some of this is dated by our standards today, it was dynamite for inflating the importance of the Yeti. Nelson concluded: "It is possible that the Yeti may prove to be an ape of a type closer to ourselves than any other existing species. [Its] description could well apply to the probable appearance of extinct Neanderthal man or to the ape-man of Africa. The Yeti could be a survivor that has lingered on long after other creatures of this type have become extinct. One wonders what he can find to live on among the snows of the Himalayas. He is said to have killed and eaten men, but this is not surprising. Our remote ancestors were cannibals just as some of our human contemporaries are."

Not surprisingly, the "Shipton foot" was becoming the most studied foot in the world. Izzard was able to share with his compatriots a surprising article by a Russian émigré scientist published in the *Manchester Guardian*: "Nature of the 'Abominable Snowman.' A Higher Form of Anthropoid?" The author, Vladimir Tschernesky, boldly disagreed with what had been reported of the Yeti's height. Based on the reported stride, his studies led him to believe that the witnesses only saw a Yeti that was not fully grown. "The length of the stride of the Snowman, as the writer of this article was kindly informed by Eric Shipton, is about 2 ft. 6 in. It is thus equal to the stride of a six-foot man. The ancestors of man, such as the Neanderthal man, or the Pithecanthropus, had relatively shorter legs and a more massive body than contemporary man. This suggests that the Snowman must be a huge size, not less than 7 ft. high. E. Shipton, who saw these mysterious footprints several times, never no-

---

[11] Izzard's reports were syndicated. Back at base camp, after weeks of sorties, members began to receive fan mail, newspaper clippings and magazine stories.

ticed anything that could be described as the marks of the hands. This indicates that the Snowman, as probably Pithecanthropus Erectus, is a completely bipedal creature."

Tschernesky concluded: "It is to be expected, according to all that has been said, that a new form of higher anthropoid will soon be discovered. This form will be either very nearly related to the fossil Gigantopithecus,[12] or it may be a living representative of this most extraordinary member of the primate family for such the 'Abominable Snowman' appears to be." [13]

Lip service to the theory of an animal Yeti was a nod to conservative wisdom. However, Mills', Nelson's and Tschernesky's words boldly albeit mutely said "missing link." Tschernesky was actually one of a number of anthropologists who were certain *Gigantopithecus* had been bipedal. In addition to a giant fossilized primate jaw found in the Silawak Hills of nearby India, large bones recovered in Africa (Tanganyika) and Java had proven to be so human they were given the name *Meganthropus*— "Great Man." Contemplated collectively, many anthropologists believed these fossils indicated a benchmark of primate evolution— manlike body with an ape-like head— and thus despite being a giant it revealed a significant stage of evolution that could cast light on our own.

As credible as these anthropologists were, their views as they applied them to Yeti nevertheless were little more of a rationalization than Hodgson, Waddell, Howard-Bury and Newman's had been with the evidence they had had before them. The difference was the Shipton photo. It opened the door for science, and with this simply more erudite Imagineering.

Tschernesky even went to the trouble of making a plaster cast of the image in the Shipton Photo and testing its function. He impressed it into snow to make what was the exact print that Shipton had photographed. What was most impressive, of course, was that Tschernesky made the foot completely realistic, with areas indicating where joints and where skin creases would be if the Yeti was truly a bipedal beast weighing a couple hundred pounds. Tschernesky's study uncannily confirmed, at the very least, the foot was ideally designed for a large primate and was suited to bipedality.

Vladimir Tschernesky has no doubt added significantly to the study of Yeti, and the Shipton footprint may indeed represent *Gi-*

---

[12] Literally "giant ape." Fossil jaw and teeth have been found in India.
[13] "Father of crypto-zoology," the famous Dr. Bernard Huevelmans, agreed completely.

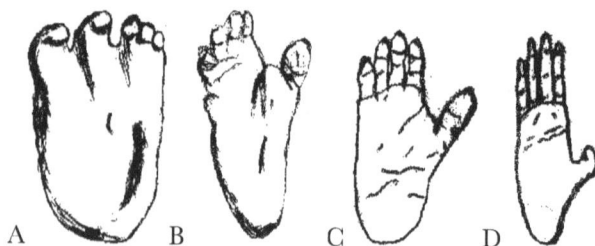

*Tschernesky followed the lead of an obvious ape-like pattern. **A**. Tschernesky's model of the Shipton Print compared to **B**. a Gorilla **C**. Chimpanzee and **D**. Orangutan. Apes essentially have 4 hands. The Shipton Print remains so different because it is a mixture of hand and foot— meant for grasping on rocky surfaces but also for bipedal walking.*

*gantopithecus*, but his experiments add little if any admissible evidence to that paltry sum that would indicate "Gigantopithecus" was some kind of missing link. In truth, the evidence for *Gigantopithecus* having once existed is really quite scant and only piecemeal assembled. Beyond that primate jaw mentioned, a few giant molars had also been found in a curio shop in China in 1934 by Ralph von Koenigswald.[14] He is actually the one who gave the name *Gigantopithecus* to them. But it is not certain whether all these artifacts truly came from the same species. Some may be from humans, others from apes. "Gigantopithecus" as a bipedal evolutionary missing link was really more a hodgepodge hope than a stable theory. Today most anthropologists would probably regard *Gigantopithecus* as a giant ape, as its name literally means. Nevertheless, the reader can understand why at the time some anthropologists *wanted* Yeti to be a giant. They wanted it to be living proof of the scattered and paltry bones in ancient sediment that they had put together in theory to represent a significant benchmark in evolution.

In the public's eye the *Daily Mail* Expedition obviously had a lot to live up to. Therefore when they never captured or filmed a Yeti, the expedition seemed all the more a failure. Izzard and Stonor could not prove Yeti was but their "Animal X," and in a "Catch 22" their vignettes only served to whet the appetites for theories of "missing links." Neither would also endorse the new popular theory that Yeti was a giant, as Tschernesky had publicly insisted, for this, too, smacked of the music comedy hall beast Britain had embellished since Howard-Bury's expedition in 1921.

---

[14] Sold as "dragon's teeth."

A. Common recreation of *Gigantopithecus*. B. Sherpa descriptions for the Yeti have never changed. If Yeti is *Gigantopithecus*, which image is more accurate?

The question is, who was right? It is a question that must be asked, even at this late date. Tschernesky brought up something hard to explain— the 2 foot 6 inch average stride of the Yeti could not be achieved unless it had manlike (and not ape-like) proportions or it was simply much bigger than the Sherpas claimed. Tschernesky opted for the bigger Yeti with short legs. It fit his view of evolution and it explained how the Yeti could go down on all fours when in a hurry. Stonor and Izzard preferred the shorter 14-year-old-boy size "Animal X" Yeti. Yet neither could explain how this version with short legs could consistently leave a long stride. If it was proportioned like an ape with short legs, Tschernesky is right: it had to be a giant. If it wasn't, it had to be proportioned like a man with long legs. Yet if that was the case, it could never go down on all fours.

The first step in the solution is accepting that there is probably more than one entity involved and that both are confused for the other. In his critique of footprints, Tilman was right. They are frequently ambiguous to interpret. What Tilman had overlooked was that the Sherpas had been consistent in describing *two* different entities— Yeti *and* the Abominable Snowman.

In fact, from the very first usage of the expression Metoh Kang-mi— the actual "Abominable Snowman"— Sherpas were eager to describe the mountain men who were "like a man, white-skinned, with long hair on head, arms and shoulders," as Tilman and

Kaulback relate. Indeed, indeed, it was the humanity of the prints Howard-Bury saw that caused the Sherpas to declare the famous words "Metoh Kangmi." The "Animal X," on the other hand, elicits "yeti" or "mih-teh" and descriptions of the cone-headed "ape."[15]

Ironically, the Abominable Snowman and the Yeti appear to always have been two separate entities; and instead of uncovering this, the *Daily Mail* Expedition unwittingly might have combined or, more likely, excluded one from the equation. Izzard in his "Yeti in Retrospect," his summation of the expedition, clarified that the expedition asked no leading questions. They only asked: "We have heard there is a beast living in your country called a Yeti. What sort of creature is it?" Perhaps this in itself was leading, limiting the Sherpas in their response to what is clearly some Anthropoid X. I say this because the Sherpas clearly told of the wild "Kangmi" for as much as 120 years before the *Daily Mail* Expedition and yet they didn't seem to insist exclusively it was an animal until the 1950s. The upshot of the *Daily Mail* expedition was to give us one mug shot. Consequently, the Kangmi has been lost for a cone-headed *Gigantopithecus.*[16]

Izzard and Stonor, however, unquestionably give us an accurate tabulation of the habits and appearance of the Yeti. Although we may dispute it is actually the Abominable Snowman, the discoveries of the 1954 *Daily Mail* Expedition are extraordinary, and they have become more so in this age of dying mythologies. Underneath the hype and hyperbole of fatuous giant ogres and albino hairy monsters they have left us a believable portrait of Pithecus X. And it is in this light that one must appreciate their contributions.

Yet there remains to be explained the actual entity behind the Metoh Kangmi. Tombazi clearly saw something human in outline, but he described a foot radically different from those of the Yeti and humans. On a couple of occasions Stonor and Izzard's hiking parties also added to this enigma when they came across different and mysterious footprints. In relating an encounter near Namche Bazar, Izzard declares that the Sherpas actually saw the hairy Mi-Teh and when coming out after it left noted that its footprint showed toes that were all the same length and even across— quite

---

[15] Even Shipton seems to have conflated them in *Menlung La* when he qualifies Tensing saw the "Yeti or Abominable Snowman." It is not clear whether Tensing equated the two or whether Shipton innocently may have done so when relaying Tensing's observations to his western readers.

[16] Aside from reporting more tracks of the Shipton style footprint both Stonor and Izzard describe finding tracks that were virtually human.

the opposite of the Shipton Photo footprint and closer to what Tombazi reported.

Unfortunately, no such print as Tombazi reported has been photographed or cast in the Himalayas. But the legitimacy of his report should not be discounted on this alone. We come again to that factor Coincidence, that tantalizing wraith that inspires, educates, and sometimes deceives so many. In Borneo and Sumatra there is said to live the *orang pendek*. It is said to be a manlike primate about 5 feet tall and described as covered with short black hair except on his head, where it is much longer. His arms are shorter than an ape, though the forearms are longer than a man's. He is very strong and he walks erect on two legs. He is quite a mystery of Indonesia even to this day. Like the Yeti his existence hangs upon some native descriptions and a few footprints.

A couple of these footprints, however, match those that Tombazi reported in the far-away Himalayas— stub toes lined up straight across and a narrow heel. Surveyor Raden Kasanredjo was the first to trace these, found near Air Masoedje. These were very similar to prints traced by Mr. Lambermon at Kaba Wetan in September 1918, both years before Tombazi's report.

Do these represent the *orang pendek* and the Metoh Kangmi? Unfortunately for any neat closure, orang pendek is an elastic term that merely means "small man." In 1915, Dr. Edward Jacobson traced a footprint near the Danau Bento Swamp, Borneo, that is also supposed to be the *orang pendek*. It is very different than those mentioned above, and very human-like. There the mystery stops, for now. The *orang pendek* has never been found despite being the subject of several searches, including a 15 year long hunt by Debbie Martyr, which ended in 2005 without finding the elusive creature.

The same, of course, can be said for Yeti. It has disappeared into the snows again and into the pages of folklore where it remains single-o the cone-headed missing link *Gigantopithecus*. Our last image

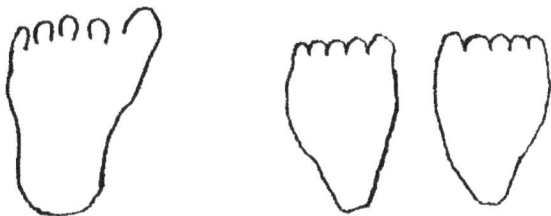

*According to published copies (Heuvelmans), left, the Jacobson print of the orang pendek; right, the orang pendek print found by Kasanredjo and Lambermon.*

*Author's copy of Tschernesky's Yeti, drawn in 1957.*

of it was not Izzard and Stonor's tempering mug shot of the Animal X. Tschernesky's image had won out. In 1959, Bernard Heuvelmans' voluminous tome *On the Track of Unknown Animals* was published to wide critical and popular acclaim. With it he became the undisputed "Father of Crypto-zoology." He completely endorsed Tschernesky's giant cone-headed ape, and through the success of his book Tschernesky's artistic recreation of the giant Yeti was cemented in the popular mind, indelibly in those who followed cryptids.

It is this Yeti that would step across the Bering Straits and land in America with only its cone-head intact. The chain of events began in British Columbia. In 1958 that province would celebrate its Centenary. For a couple of years much talk had been going around about what cultural symbols should be used. Naturally, the Sasquatch was one of them, that is, J.W. Burns' incarnation of the primitive Indian. A skeptical local newsman, John Green, had been studying the possibility that the Sasquatch could actually be real. But he found it hard to believe it could possibly be a primitive Indian. He found it even harder to believe it could be a Yeti, and yet this was the opinion of the first person he had ever met who took the legend seriously.

In 1956, an enthusiastic Swiss immigrant named René Dahinden had come to Green's newspaper and enquired about the Indian stories, telling Green he had already collected as much library information as he could on the subject. He was certain Sasquatch was a Yeti. In his own writings later, Dahinden would make it abundantly clear that he was set to this course by the news announcements in 1954 of the *Daily Mail* Expedition. Living in Alberta at that time, Dahinden was captivated when his boss, Wilbur Willick, replied to the broadcast that "them things" also lived in British Columbia. By the time Dahinden had moved to the west coast he had added Izzard's *Abominable Snowman Adventure* to his collection and Tschernesky's Yeti giant to his theorizing.

Yet despite Dahinden's certainty, Green greeted the young Switzer's theory with lackluster dismissal, telling him it was all Indian myth. But in 1957 Green wasn't so certain anymore. With all the build-up to the celebrations, the papers were running stories. One immediately caught his attention. A hunter named William Roe reported he had seen what he thought to be a Sasquatch up Mica Mountain in the Canadian Rockies in October of 1955. Yet what he described was more animal than human, more in step with the notion of the Yeti. Green was soon in contact with him and asked for a detailed description, even imposing enough to ask for a legal affidavit.

William Roe complied. The Roe encounter is one of the most authentic there is. The Roe sighting is to sightings what the Shipton footprint is to footprints and it as significant in finally destroying White Man's belief Sasquatches were giant people. The legal document, dated August 26, 1957, begins:

Exhibit A

Ever since I was a small boy back in the forests of Michigan, I have studied the lives and habits of wild animals. Later, when I supported my family in northern Alberta by hunting and trapping, I spent many hours just observing the wild things. They fascinated me. But the most incredible experience I ever had with a wild creature occurred near a little place called Tete Jaune Cache, B.C., about 80 miles west of Jasper, Alberta.

I had been working on the highway near this place, Tete Jaune Cache, for about two years. In October, 1955, I decided to climb 5 miles[17] up Mica Mountain to an old deserted mine, just for something to do. I came in sight of the

---

[17] This must be a typo. Mt. Everest is only 5 ½ miles high. He must mean 5,000 feet or the hike was five miles from where he parked his car.

mine about three o'clock in the afternoon after an easy climb. I had just come out of a patch of low brush into a clearing, when I saw what I thought was a grizzly bear in the brush on the other side. I had shot a grizzly near that spot the year before. This one was only about 75 yards away, but I didn't want to shoot it, for I had no way of getting it out. So I sat down on a small rock and watched, my rifle in my hands.

I could just see part of the animal's head and the top of one shoulder. A moment later it raised up and stepped out into the opening. Then I saw that it wasn't a bear.

This, to the best of my recollection, is what the creature looked like and how it acted as it came across the clearing directly towards me. My first impression was of a huge man about 6 feet tall, about 3 feet wide, and probably weighing somewhere near 300 pounds. It was covered from head to foot with dark brown, silver-tipped hair. But as it came closer I saw by its breasts that it was female.

And yet, its torso was not curved like a female's. Its broad frame was straight from shoulder to hip. Its arms were much thicker than a man's arms, and longer, reaching almost to its knees. Its feet were broader proportionately than a man's, about 5 inches wide in front and tapering to much thinner heels. When it walked it placed the heel of its foot down first, and I could see the gray brown skin or hide on the soles of its feet.

It came to the edge of the brush I was hiding in, within 20 feet of me, and squatted down on its haunches. Reaching out its hands it pulled the branches of bushes toward it and stripped the leaves with its teeth. Its lips curled flexibly around the leaves as it ate. I was close enough to see that its teeth were white and even.

The shape of the creature's head somewhat resembled a Negro's. The head was higher at the back than at the front. The nose was broad and flat. The lips and chin protruded further than its nose. But the hair that covered it, leaving bare only the parts of the face around the mouth, nose and ears, made it resemble an animal as much as a human. None of its hair, even on the back of its head, was longer than an inch, and that on its face much shorter. Its ears were shaped like a human's ears. But it eyes were small and black like a bear's. And its neck also was unhuman, thicker and shorter than any man's I had ever seen.

As I watched this creature I wondered if some movie company was making a film in this place and that what I saw was an actor made up to look partly human and partly animal. But as I observed it more I decided it would be impossible to

*Based on Myrtle Walton's (Roe's daughter) drawing (under his supervision) of what he saw. Roe gives the creature a higher inseam and more human proportions than an ape.*

fake such a specimen. Anyway, I learned later that there was no such company near that area. Nor, in fact, did anyone live up Mica Mountain, according to the people who lived in Tete Jaune Cache.

Finally, the wild thing must have got my scent, for it looked directly at me through an opening in the brush. A look of amazement crossed its face. It looked so comical at that moment I had to grin. Still in a crouched position, it backed up three or four short steps, and straightened up to its full height and started to walk rapidly back the way it had come. For a moment it watched me over its shoulder as it went, not exactly afraid, but as though it wanted no contact with anything strange. The thought came to me that if I shot it I would possibly have a specimen of great interest to scientists the world over. I had heard stories about the Sasquatch, the giant hairy Indians that live in the legend of the Indians of British Columbia, and also, many claim are still, in fact, alive today. Maybe this was the Sasquatch, I told myself.

I levelled my rifle. The creature was still walking rapidly away, again turning its head to look in my direction. I lowered the rifle. Although I have called the creature 'it,' I felt now that it was a human being, and I knew I would never forgive myself if I killed it.

Just as it came to the other patch of brush it threw back its head and made a peculiar noise that seemed to be half laugh and half language, and which I could only describe as a kind of a whinny. Then it walked from the small brush into a stand of lodge-pole pines.

I stepped out into the opening and looked across a small ridge just beyond the pine to see if I could see it again. It came out on the ridge a couple of hundred yards away from me, tipped its head back again, and again emitted the only sound I had heard it make, but what this half-laugh, half-language was meant to convey I do not know. It disappeared then, and I never saw it again.

I wanted to find out if it lived on vegetation entirely or ate meat as well, so I went down and looked for signs. I found it in five different places, and although I examined it thoroughly, I could find no hair or shells or bugs or insects. So I believe it was strictly a vegetarian.

I found one place where it had slept for a couple of nights under a tree. Now, the nights were cool up the mountain, at this time of year especially, and yet it had not used a fire. I found no signs that it possessed even the simplest of tools. Nor did I find any signs that it had a single companion while in this place.

Whether this creature was a Sasquatch I do not know. It will always remain a mystery to me unless another one is found.

I hereby declare the above statement to be in every part true, to the best of my powers of observation and recollection.

It seems unquestionable that Roe saw what he saw. But instead of replacing White Man's impression of the giant mountain ogres an interesting phenomenon of pop culture would slowly take its place: Bigfoot, the American Yeti. A false veneer was being replaced by an even greater false veneer. Something was definitely afoot in the Pacific Northwest that was neither a charming Indian myth nor an Eurasia giant ape.

# CHAPTER 3

## Something's Afoot

A few years before worldwide Yeti popularity the Saskahaua district had bid farewell to Indian agent J.W. Burns, who had now retired and moved on to San Francisco. The popularity of the Sasquatch had waned since his departure and only a few landmarks recalled the yearly "Sasquatch Days" festival. As a result there was no one in place capable of going to the Indians in an attempt to see if an unknown primate would better tally with their descriptions, at least for one of the tribes, or even hand popular sketches to them of Yeti to see if it would spark a reaction or to see how they might adjust it. Had this been done, the reaction of the Indians would have been significant; either there would have been a shake of the head or an impressive composite.

With hindsight one can say this was the disastrous moment for revealing Sasquatch reality. The crucial juncture had come to dispel White Man's impression of the "Indian myth" of giant Indians. But the only newsman interested, John Green, was far outside of being in any confidence with the Indians. He was a relative newcomer to the Saskahaua, and he was, like all Whites, only acquainted with the hybrid White Man version of the legend. And his only encounter with a believer so far was with Dahinden, a man who firmly believed Sasquatch was a Yeti. Although all this had contributed to his silent questioning of the old image, none of it really gave him an alternate choice. Soon, however, he would have his own unpublicized angle on it that would convince him Sasquatch was real.

Due to the Centenary "the subject of sasquatches tended to come up in many conversations," Green confessed. It was from an acquaintance of his, Martha Kirkman, that Green first heard an intriguing story about Kirkman's cousin, Jeannie Chapman. After hearing the story, Green was able to talk to the Chapmans himself.

The Ruby Creek Incident, as it later became known, is one of the most significant but, paradoxically, glossed over cases. It occurred near the Fraser River in British Columbia on October 21, 1941. The small cabin belonged to George and Jeannie Chapman and their three children. Something unusual was first seen by the daughter, little Rosie, who was playing in the garden. She rushed into the house and said "Mommy, a big cow is coming out of the woods." Mrs. Chapman went to have a look and was absolutely terrified by what she saw. She told John Green that it was a tall hairy man walking across the field coming toward the house. Judging by field posts, she estimated it was 7 and a half feet tall.[18]

Believing it was indeed a Sasquatch, Jeannie Chapman immediately gathered her children and led them away to the river, keeping the cabin between them and the creature until they made it down to the bank, which they then followed until they made it to safety at another farm.

George came home and was shocked to see that there were huge footprints all over the place and that his wife and children were missing. He also found that a barrel of salt fish had been broken into and some of the fish were half-eaten and scattered about before the shed where it was kept. It would have taken a powerful force to have done this, and he was now doubly worried over his family. Along the creek, however, he found their tracks. After following them he found them and heard the story straight from his wife.

Mrs. Chapman's mistake perhaps was in calling the creature a Sasquatch. Her report caused such a furor amongst the Indians that even the Vancouver *Province* ran an article explaining it away as a 10 foot bear— "one of the largest ever." Sasquatch, of course, at the time carried the connotation of a giant primitive Indian. Worse, this was still within the memory of the 1934 spate of Sasquatch ter-

---

[18] Jeannie later told naturalist Ivan T. Sanderson that the Sasquatch had a noticeably small head for its body size.

ror at Harrison Hills that earned national news in both Canada and America and then national disinterest when no Sasquatch man was apprehended. No one wanted to repeat that debacle. Desiring to avoid this may have made the reporter giggle this thing off as merely a simple Indian imagining the legendary Sasquatch "giant Indians" had come to life.

In any case, Jeannie Chapman was terror stricken by the encounter. By the end of the week after the "thing" had returned several times at night she refused to remain at the farm. George had no choice but to abandon it, and the little house remained empty and dilapidated until it fell apart, and the land fallow until it was reforested.

After hearing the stories (and uncovering the shabby article) Green backtracked and found more impressive information on the incident. He discovered that two men hadn't dismissed the case. One, Esse Tyfting, he already knew. At the time of the incident Tyfting had lived at Ruby Creek (it is only 12 miles up the Fraser River from Agassiz). Along with Joe Dunn, a deputy sheriff from Washington State, Tyfting had gone to the farm to investigate. By 1957, Dunn had already passed away, but Green was able to confirm Tyfting's account from Dunn's son, who had kept his father's report and even a tracing of one of the footprints.

According to both Dunn's and Tyfting's accounts, the footprints traversed the field coming from the forest, and in the potato patch they were 2 inches deep, crushing the buried potatoes below. The footprints then circled the house and then went to the outshed where the "thing" broke into the heavy barrel of fish. There was some disagreement in Dunn's report and the recollection of Tyfting whether the heavy barrel had been lifted and broken open by bashing it against the wall or just busted into. No matter which, half-chewed fish were lying about. The footprints then headed to the creek, where, Green deduced, the creature must have washed the salt from the fish off its lips.

The footprints then went back into the forest. Here one of the most remarkable things was discovered. There was a Canadian Pacific Railway fence at the edge of the property. Unbelievably, the footprints continued on both sides of it as if the fence hadn't been there, one print on this side and the other on the other side. The creature had simply *stepped over it,* then walked across the tracks and up a mountain.

The CPR fence measured 43 inches tall. If the Sasquatch was 7 and a half feet tall, it meant that its legs were indeed about half its overall body height (7.6 feet is 90 inches), the proportions of a man.

Green was so careful in investigating this case (the first real case ever investigated) that he asked Tyfting to be interrogated before giving his declaration. Present were Green and the local magistrate, Lt. Col. A.M. Naismith, a trial lawyer of some repute. Here is just a brief extract from the questioning which appears in Green's self-published *On the Track of the Sasquatch*:

Q— How big were the tracks?

A-- About 16 inches long, 4 inches at the heel and 8 inches across the ball of the foot.

Q— Were there five toes?

A-- Yes, but no claw marks. . .the stride between the prints was four feet between the heel and toe, all through the potato patch.

Q— Did you measure the footprints?

A-- Yes.

Q— You, yourself?

A-- Yes. I measured them and after a man came from across the line [Deputy Sheriff Dunn] we measured them again.

Q— And what did Mrs. Chapman say about the Sasquatch?

A-- She said it was a big hairy man.

Q— And what condition was she in?

A-- Scared to death.

Tyfting also testified that the deep footprints could only have been made by a creature weighing from 800 to 1,000 pounds. Tyfting was also very familiar with bears. These were human-like footprints, and the creature walked on its hind legs the entire time. Green was able to verify this for himself from Dunn's son, who also let Green trace the tracing. This footprint is amazing. The tracing is that of a giant flat foot. It is humanoid, but not human; it is broad at the ball of the foot and narrow at the heel. Unlike a human foot, however, the five toes are almost even across.

This tracing becomes Exhibit B of evidence. Fortunately because of its preservation we have a clear and reliable outline of the fabled Sasquatch's footprint, at least for one of the "tribes." For the Sasquatch it is as significant as the Shipton Photo is for Yeti. In addition, we can compare it to William Roe's account to find it matches

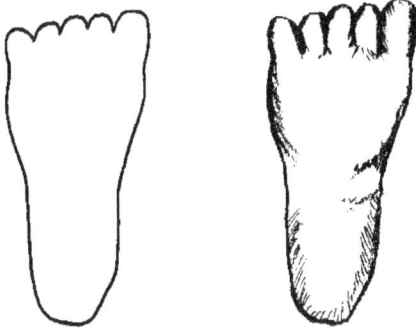

*Exhibit B of evidence. John Green obtained a copy of the tracing from Dunn's son, from which the above drawing on the left is based. Right, is a drawing of what the foot might have looked like. It is clearly not human or a human variation.*

identically with what he saw in 1955. An account of such a footprint also exists in J.W. Burns' very first article. Indian brave Peter Williams had a frightening encounter in which the Sasquatch leapt from a rock and chased him across the river to his cabin. Williams bolted himself in and the Sasquatch, growling and howling, tried to get in, shaking the cabin so bad it was completely sprung (Burns provides a picture of the tattered cabin). Next morning, Williams came out and saw the footprints. "The tracks measured 22 inches in length, but narrow in proportion to their length."

A similar foot is described in an obscure report made as far back as 1818 in New York State. The article appeared in the September 22 edition of the *Exeter Watchman*. The report was dated September 6 at Sacket's Harbor.

Report says, that in the vicinity of Ellisburgh, was seen on the 30th Ult by a gentleman of unquestionable veracity, an animal resembling the Wild Man of the Woods. It is stated that he came from the woods within a few rods of this gentleman,—that he stood and looked at him and then took his flight in a direction which gave a perfect view of him for some time. He is described as bending forward when running— hairy, and the heel of the foot narrow, spreading at the toes. Hundreds of persons have been in pursuit for several days, but nothing further is heard or seen.

One can debate if what Dunn traced, what Roe described, and what this 1818 report declares, all come from the same species. But one thing is *certain*— they *do not* describe the Yeti foot at all. This does not match the Shipton Print in any way. In short, Dunn's print scuttles the whole idea about the Sasquatch merely being an American version of the Yeti.

Exhibits A and B, the only two reliable footprints so far, would have established a sure foundation for future interpretations of footprints. Each would have allowed researchers to pursue entirely separate junctures without mixing them. Each on its own could also be used to explain, validate or invalidate the number of nebulously described footprints in old newspaper articles or, even better, those that would soon come to be. Abiding by this criterion would have spared researchers a terrible tangent that would warp the pursuit of Sasquatch until the present.

In California, 1958, the word Sasquatch inspired nothing but a vacuous look or at best it might have evoked a vague recollection of some Canadian fairy tale. Nobody also would have known what to make of huge footprints found in the wilds. Thus it is not surprising that in that very year when 16 inch footprints were found on the logging roads of Bluff Creek in the Klamath Mountains the "miscreant" was summed up as nothing more than "Big Foot." The first incident got widespread attention, beginning with an AP article, dateline Eureka.

> Jerry Crew, a hard-eyed catskinner who bulldozes logging roads for a living, came to town this weekend with a plaster cast of a footprint.
>
> The footprint looks human, but it is 16 inches long, 7 inches wide, the great weight of the creature that made it sank the print 2 inches into the dirt.
>
> Crew says an ordinary foot will penetrate that dirt only half an inch.
>
> "I've seen hundreds of these footprints in the past few weeks," said Crew.
>
> He added he made the cast of a print in dirt he had bulldozed Friday in the logging operation in the forest above Weitchpeg, 50 miles north and a bit east of here in the Klamath River country of northwestern California.
>
> Crew said he and his fellow workmen never have seen the creature, but often have had a sense of being watched as they worked in the tall timber.

BIGFOOT, as the Bluff Creek people call the creature, apparently travels on-ly at night.

Crew said he seems fascinated by logging operations, particularly the earthmoving that Crew does with his bulldozer in hacking out new logging trails.

"Every morning we find his footprints in the fresh earth we've moved the day before," Crew said.

CREW SAID Robert Titmus, a taxidermist from Redding, studied the tracks and said they were not made by any known animals.

"And they can't be made by a bear, as there are no claw marks."

"The foot has five stubby toes and the stride averages about 50 inches when he's walking and goes up to 10 feet when he's running."

TWO YEARS AGO reports from this area told of logging camp equipment tumbled about, including full 50-gallon drums of gasoline scattered by some un-known agency.

Well, there it was. "Big Foot" was coined. Holding his odd plas-ter, Jerry Crew had been dazzled in the flashbulbs of the *Humboldt Times*. Hundreds of newspapers carried the pic-tures of him proudly proffering that big foot, in-cluding the Vancouver *Province* which boldly stated in headline "New Sasquatch Found." But what is it? What he holds is not the Sasquatch print or the Yeti print. It is quite obviously a plas-ter cast of a flat enlarged human foot.

In addition to this curious footprint another set of tracks had appeared at the same time. Alt-hough they find no mention in the article, Bill Chambers, the reporter for the *Humboldt Times*, was shown both tracks in order to validate the story. The other set of tracks had been found by Crew's boss, contractor Ray Wallace. This set of tracks had been made by a foot even stranger

Crew Cast, 1958

than those that made the Crew Print. They were shaped like an hourglass and had an unexplained groove in the ball of the foot.

John Green had read the *Province* article and by late October had driven down and talked to Jerry Crew and another lumberman, Jess Bemis. Both then took him up the logging road and showed him some weathered prints. Green had also heard a number of rumors how it was all a hoax. However, he decided to dig further. He drove to Redding and talked to Bob Titmus. Titmus confessed that he had been told by hunters for years about giant footprints in the woods but he always assumed they must be bears. When Crew reported these prints, he finally drove to Bluff Creek and had a look. He simply could not explain them.

Titmus promised to stay in touch as matters at Bluff Creek unfolded. Two weeks later in November he was good to his word and wrote Green that more footprints had turned up on a sandbar. Green quickly returned to California, and both went out to examine the tracks. Pressed into clean sand it was plain to see these were made by that strange hourglass foot with an unusual groove in the ball. (It is not certain Green and Titmus had known that there had been two different types of footprints involved initially.) Green was indeed impressed at what he saw. He would later write in *On the Track of the Sasquatch*: "For the first time I had a chance to appreciate the tremendous pressure with which the prints are made. Where they sank an inch into the sand my boots made only a heel print and a slightly flattened area at the center of the soles. To make a hole an inch deep I had to jump off a log about 2 feet high and land on one heel."

Green was clearly taken by the enigma before him. His investigations in Canada had convinced him Sasquatch, whatever it was, was indeed real. Now, in California, he was faced with this. Bringing him even closer to a decision, he then learned from Betty Allen "a newspaper correspondent who lived in the same town as Jerry Crew that Indians in the area knew there were big hair covered people in the remote valleys." Green added: "But there had been

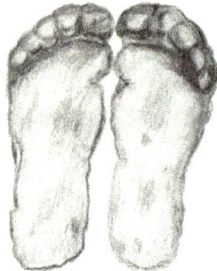

*The 1958 Sandbar prints. They are markedly different from the Crew print.*

no equivalent of Mr. Burns to tell the white community the story. Those who were prepared to accept that the big footprints were real had at first no explanation for them."

However, it was Green who now supplied the answer: Sasquatch. The same creatures said to exist in far away British Columbia had to be the "Big Foot" of California. The domino effect has seldom been clearer. The connection had to be made by a Canadian or else the big prints of Bluff Creek merely would have been funny large human prints and nobody would have known what to envision as their maker.

The connection is nonetheless a dubious one. It is obvious that the Crew Print and the Sandbar Print are themselves from radically different feet. More significantly, it is also obvious that neither bear a remote likeness to the Ruby Creek Print. Why John Green made the connection is hard to say. The Vancouver *Province* might be forgiven for running the October 6 AP bulletin under the somewhat wistful heading "New Sasquatch Found" above a picture of Crew holding the large flat human foot, but Green had the advantage of having seen and then traced the Ruby Creek print.

Perhaps I'm avoiding the issue. As we now approach the "era of Bigfootery," the era that created the cloudy window through which most people view, accept or reject Sasquatch, it is best to begin here with the task of putting "Bigfootery"— that redoubtable pursuit of Bigfoot— and its chief contributors in their actual light. I would prefer not to do this. Rather I would prefer to have the luxury Bigfooters had 50 years ago of simply following up on and leading people through the investigation of something exciting and new. At the very least Bigfooters could do this by collecting a potpourri of interesting old stories to justify their pursuit of what was skeptically greeted as modern myth. But as time and Bigfoot's popularity increased, Bigfootery uncritically assimilated any number of evolving, contradictory and odd claims that created a creature wholly different than the "animal human" that had existed prior to Bluff Creek and Bigfoot's media existence. Fifty-years of water has gone under the bridge, and I must critique this popular but terribly inaccurate and carnival episode in order to expose a false image. I cannot ignore the error for delicacy sake.

The initial excuse to explain John Green's error can only go so far. And this must have been, to put it politely, that he must have deeply wanted to believe in Sasquatch and that Bluff Creek provided the only tangible *current* evidence so far in his personal quest. This self-deception may have been short lived had Bob Titmus not

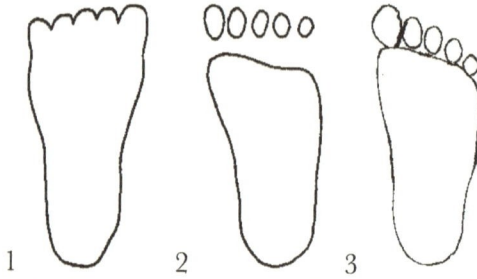

1) *The Ruby Creek Print tracing drawn by Dunn and copied by Green.* 2) *Bob Titmus' tracing of the Crew Print. (3) The actual toe arrangement on Jerry Crew's plaster cast.*

come to play a very decisive part. Green had never seen a clear image of the Crew Print. He had only seen the picture in the *Province* (which, frankly, should have been enough). But Green *had shown* Titmus his tracing of the Ruby Creek Print. Soon after this, Titmus would give Green his tracing of the Crew Print for comparison.

To put it bluntly, Titmus' illustration is phony. Its purpose seems to imitate the Ruby Creek Print and therewith suggest a pattern. Yet instead of exposing Titmus' blatant artistic fib, Green was wholeheartedly convinced it was genuine. Ten years later in his first book, *On the Track of the Sasquatch*, he used the comparison to anchor his equation of Bigfoot as Sasquatch. "What particularly impressed me," wrote Green, "was the similarity between the outline of these tracks and the tracing I had of one of the Ruby Creek footprints. They did not diverge by more than half an inch at any point, and could easily have been made by the same individual."[19]

Such a phenomenal statement should even make Ripley sit up and do a doubletake. There was, in fact, no resemblance other than the one Titmus wished to fabricate. And Green had more than 10 years to confirm or expose the accuracy of Titmus' drawing. Many newspapers printed and reprinted the pictures of Crew holding his famous cast of the 16 inch Bigfoot foot. And, indeed, it appeared in many of Green's own books. Green is even pictured holding the cast in the Willow Creek Museum. Yet in almost every publication of Green's various booklets/books, including recent reprints, and many pivotal documentaries thanks to Green's assistance, the bogus comparison is promoted.

Moreover, Green's deductions do not incorporate any explanation for those odd hourglass footprints, the only actual footprints he

---

[19] *On the Track of the Sasquatch*, page 13.

ever saw *in situ.* Yet they were the prints which had convinced him Bigfoot was real. On top of such analytical gaffs, Green never questioned any of Bob Titmus' future contributions to Bigfootery. And yet Titmus, sadly, would become one of the most key (and lifelong) original "Bigfooters" and a source of repeated and often what appears to be intentionally fabricated error. This propensity became apparent to René Dahinden, one of the other key founding Bigfooters. Titmus was, to use his words, "a fucking storyteller." The tragedy of this could have been minimized by careful investigators. But it was actually compounded by the fact that Green, who would almost become the publicist for Bigfoot via his many booklets and newspaper articles, relied completely, almost naively, on Bob Titmus. Because of this the machinations of one Bigfooter became the weak heel upon which Bigfoot would walk around the world. Even to this day Titmus' false comparison underscores for the uninitiated that Bluff Creek's Bigfoot is intimately associated with the dreaded giant hairy Indian of British Columbia. This was the pivotal mistake that started the snowball of misconceptions.

This snowball and all its growing misconceptions *would be* Bigfootery. It is the era of which we are still technically a part. This one snowball by Titmus was the seed, but it could not have been set in motion without those who would become Bigfooters. Amongst the core of the early Bigfooters the Bigfoot/Sasquatch concept was secured. And, sadly, the method of investigation and analysis that created and accepted the Titmus tracing comparisons, that moreover also didn't notice the monumental difference in the Crew, Hourglass (Sandbar) and Yeti prints, would dominate Bigfootery's greatest researchers.

I'm not sure whether it is better for the reader to block out knowledge of this or to keep it in the forefront of your mind as we continue. If you did the former the futility of Bigfootery's discoveries and indeed its whole era would eventually dominate your senses. If doing the latter, the irony of Bigfootery and its era will certainly be accentuated, which I feel is more appropriate. But I leave it to the reader to size up the era that now follows. It must be placed in context, whether irony or futility be its ambiance.

The Pacific Northwest Expedition, begun in 1959 as a result of the publicity accorded the Crew Print, was funded by Texas millionaire and oilman, Tom Slick.[20] This organization was the very first and perhaps to this day the most serious and organized concerted effort to track down Sasquatch. Founding members were: Slick, John Green, René Dahinden, Ed Patrick, S. Kirk Johnson, Bob Titmus and even Slick's secretary, Gerri Walsh. These were, of course, the core group of people who expressed early interest in the subject. Soon to join would be Ivan Marx, Ivan Sanderson and eventually Peter Byrne.

"Track down Sasquatch" is actually an ironic statement; indeed it is in substance an inaccurate statement. Thanks to the Titmus/Green misperception Slick's expedition was largely limited to northern California and Bigfoot. Bob Titmus was the deputy leader, lending the "expert tracker" element to it; for the better part of 3 years following up on track reports, usually in the Bluff Creek area. Ivan Sanderson, being a well-known and published naturalist, F.L.S., F.R.G.S., F.Z.S., and so forth no less, added a "scientific" respectability to it from afar. S. Kirk Johnson was the associate leader, another Texan who had headed Slick's Himalayan expedition for the Yeti a couple of years before.

Via the Canadian contingent much of Sasquatch and its contradictory attributes (human and animal), as reported by Roe and Indian tradition, crept into the Bigfoot makeup. Added to this confusion was the extraordinary belief of the group that the Yeti and Sasquatch must be one and the same. The aggressive nature of the giant hairy Indians, as contained in the early accounts, was fertile material to bolster Ivan Sanderson's "caveman" approach (don't forget Bigfoot also had human feet!). Sanderson, who quickly became PNE's scientific advisor, wrote his early book *Abominable Snowmen: Legend Come to Life* on the subject in 1961 with a subtitle that said it all: "The story of sub-humans on five continents from the early ice age until today."

This voluminous tome gives one an almost cringing insight into the strange impenetrable illogic of PNE, for it is certain that on similar data John Green's opinion leaned toward Sasquatch as a Yeti and therefore as *Gigantopithecus*; and Dahinden retained a decidedly sanguine approach to the subject, being always ready to shoot one to find out.

As a happy and profitable median, John Green maintained a prolific output of chronicling every report, a common ground where pathetic human trog, vicious attacker, and placid giant ape could

---

[20] Originally known informally Northwestern and Northwestern Pacific.

find longevity. Out of all the spine tingling reports the *Humboldt Times* dished out, Green selected to bring forward in his own publications a story how one construction worker who slept in an old shack on Bluff Creek Road (which was a gravel road, the main road at the time) heard something moving around outside. When he yanked the door open he was face-to-face with a "man-like giant." In his panic he picked up a chocolate bar and handed it to the "Bigfoot," which took it and then departed contented.

News about Bigfoot unquestionably had the *Humboldt Times* in Eureka as its major outlet. After the success of having broken the Jerry Crew story, the *Times* uncritically covered every whisper that came out of the logging area. Bill Chambers, their roving reporter, was actually the first reporter to have been shown the prints *in situ*, guided by Crew's boss, Ray Wallace.[21]

There was no question that most of the Bigfoot incidents surrounded Wallace's Granite Construction Company and its work bulldozing roads. It wasn't long therefore before Ray and his brother Wilbur became the source of some of the most exciting information. They were main cogs in the wheel of the legend of moved full 55 gallon drums of diesel fuel, 250 pound "earthmover" wheels rolled a ¼ of a mile down roads, and thrown galvanized culverts. Ray was even quoted in the October 14, 1958, edition as saying that he measured the running stride of a Bigfoot and it measured 10 feet!

Soon the stories began to spread beyond the brothers Wallace. One chilling story was that of mutilated dogs found south of Eureka, "their bodies still warm." One had supposedly been slammed against a tree. The *Humboldt Times* reported on October 19, 1958: "NEW BLUFF CREEK MYSTERY PUZZLES INDIAN: 4 DOGS FOUND RIPPED TO PIECES." This was declared to be the sign of a Bigfoot "temper fit." The story is second hand at best, coming through Harold C. Goodwin relating what his Indian worker Curtis Mitchell had reported. Goodwin himself had been a resident of Humboldt County since 1898, during the era of the "Kangaroo Man," and had long heard since childhood of the "giant footprints." He had, like so many others, dismissed them as just "a joke." Now he wasn't so sure. The discovery of the mutilated dogs "is when we old timers start to believe."

---

[21] A December 6, 1965, San Francisco *Chronicle* article confirms this: "Bill Chambers, at that time a reporter for the *Humboldt Times*, inspected the prints found by Crew and another set of giant prints found in the Bluff Creek area by contractor Ray Wallace."

The *Humboldt Times* had so much success that even the San Francisco *Chronicle* tried its hand with Bigfoot. It rooted out independent witnesses, including one hunter, an O.T. Edwards of Fresno, and his story of coming across such a "hairy man animal" during WWII while hunting in the Siskiyous. He was called to its attention by the "damnedest whistling-scream that I had ever heard from right behind me." The *Humboldt Times* reporter, Betty Allen, catskinner Jerry Crew, and Al Hodson, a local businessman, were all portrayed as interviewing witnesses who had "heard their eerie howls" in order to get to the bottom of just what this Bigfoot really was.

From there it went nationwide. Ray Wallace's solid manner and hard-hitting attitude made him the preeminent picture of source-reliability. With the *Humboldt Times'* referral, Ray and Wilbur were the source behind much of Ivan Sanderson's famous 1959 article for Fawcett's *True Magazine*. "The Strange Story of America's Abominable Snowman" was read by millions. Sanderson built up Ray Wallace as "a hard-boiled and pragmatic man" and underscored how worried Ray and his brothers were about Bigfoot inasmuch as it was hard to keep some men on the job. "Wallace was convinced," wrote Sanderson, "that somebody was trying to disrupt his work, and this made him furious."

Sanderson reported to the nation that Ray Wallace was so mad about the job repercussions that he had hired Ray Kerr to investigate. Kerr was not the disinterested security investigator, however. Kerr had heard about what was going on and actually approached Wallace for a job so he could be on site and investigate the sensational giant. This Wallace did. And it wouldn't be long before dear Ray Kerr had an extraordinary encounter that had found its way into the *Humboldt Times* on October 14, 1958, before it caused nationwide thrills in *True* in December 1959. A "bounding Big Foot" ran across the road and Kerr estimated the hairy thing must have been 8 to a whopping 10 feet tall!

Sanderson' article got some national yuks in *True*'s March edition, 1960, but Bigfoot still made local news thereafter. A report of those strange hourglass prints was made in August 1962 when 3 young men, including 2 midshipmen on vacation from Annapolis, spotted them in the creek bed by one of Wallace's lumber roads. This was along the Klamath River about 36 miles north of Weitchpeg, northeast of Onion Mountain on the Siskiyou County border. The astute reporter for the ever-alert *Humboldt Times* reported for August 10, 1902, that the young men "rekindled interest in the mysterious monster of the Sixes Rivers wilderness. . ." With that one comment the perspicuitive newspaper reporter paved the way for the famous flap of incidents that seized Bluff Creek in 1963. "Flap" is the only word to describe the incidents that reportedly went on. Throughout 1963 the *Humboldt Times* did follow-up stories, fueling the fears of the old "Kangaroo Man" by noting how lumber workers continued to find huge tracks on the logging roads. Another story had it that when the lead crew was bulldozing the road they heard a crash far behind them and thought it was the culvert crew. When they went back to investigate, they found that the parked trailer had been thrown upside down, still full of its load of culverts. It had simply been flipped over. It took 12 men to right it again. Government road inspector Pat Graves commented that "Bigfoot seemed to have a mad on for culverts."

Some men even claimed that a culvert was found in a stream at the bottom of the road embankment. The lack of any damage to the undergrowth on the intervening slope meant "that they went all the way in the air." In other words, immense strength had to be used to lift the culvert and then hurl it with such force that it didn't fall down the slope; it just flew in mid-air until it hit the stream. The logging foreman, Dave Blake, even confessed it was hard to keep men on the job because they were so unnerved.

This was the legend of Bigfoot. It opened the forests of our humdrum lives and gave us the scientific discovery of the century in our own backyards. He was ours 100 percent— ornery but harmless, dreadful yet intriguing. Ironically, despite PNE's founding heralded by the bold headline "Slick Thinks Bigfoot Kin of Abominable Snowman," Bigfoot remained a mélange. The image of the cone-headed Yeti had not yet won out. Nothing, in fact, had changed by 1965 when the San Francisco *Chronicle*, that venerable "voice of the West," ran a series of stories by George Draper starting on December 6. They were delightful articles that capitalized on the old Indian stories and frontier journals of the "animal humans," and it was even appropriately entitled "Animal-Men of the Northwest." It provided us with a stalwart image of the 1960's father/hunter in fedora compared to the whopping 10 foot tall troglodyte "Man Animal." It was pure Americana, rooted in the giant human tracks at Bluff Creek. Bigfoot was Neanderthal, ogre, and caveman. In addition, John Green had also published the account of a Canadian named Albert Ostman in his own *Advance*, which spoke of this trapper's capture and stay with a family of Sasquatches. Each was modeled on some caveman, and the father Sasquatch could even

*The San Francisco Chronicle gave the West its last image of the frontier "animal human" in the above comparison.*

grunt exhilarating words like "soka, soka" and "ook."

Like Sanderson's, Green's writings were a melting pot of uncritical journalism. As such there was no way their works could give us one undeniable image. The paradox is that we kept our old single albeit vague impression of the giant, hairy primitive human. Both Green and Sanderson were also a true reflection of the substance of PNE. Green was a brilliant chronicler but not analyzer, and Sanderson was an agenda-driven, often flippant theorist.

Another reason perhaps why the Yeti image had not yet won out was that its chief protagonist— René Dahinden— quickly grew repulsed at PNE and left. Thanks to his new friendship with Green, he had joined PNE in November 1959, receiving 350 bucks and living expenses a month from the Texas millionaire. However, Dahinden saw how odd the entire expedition was run. In one case, Slick gave a blank check to a couple that merely reported footprints, enlisting them to become trackers. He soon had to take them to court to stop the drain on his account. Titmus also proved to be too bizarre for words. One day he appeared at camp with a collection of used tampons, which he confessed he had stolen at gas station public bathrooms. He had eventually amassed quite a number and nailed them to trees. His theory was that Bigfoot's libido would be attracted by the scent and then they'd have him captured. At the end of January 1960, Dahinden had left the group.

It ambled along until 1962, its rag-tag members scouring the area of Bluff Creek for the giant with enlarged human feet. In its 3 years it had chased a carnival creature. The result was a carnival dossier as logical as a group with such differing (and contradictory) *weltanschauungs* as Dahinden's and Sanderson's could make it. Its members more than once traveled long distances to take witness depositions concerning "unknown animals" and then take plaster casts of enlarged human footprints— footprints that were as believable as the chocolate bar and many other *Humboldt Times* stories. PNE did not intend to become the center of bizarre reports, but considering contradictory motives— such as fame, money, glory and radically opposite points of view— objective analysis did not happen.

It is rather sketchy whether the inability of some of the key members to maintain a steady field presence was responsible for Slick calling in Peter Byrne or whether Slick was becoming disenchanted with some of the PNE members. A more likely reason would be a combination of both. Tom Slick no doubt realized he had roped into his organization some dead weight. He cut checks for a number of the Expedition, but the upshot so far was contradictory plaster casts, loads of moose dung and some hair samples. Dung and hair proved useless, since biologists made it clear one needs to compare it to known hair or dung from a Bigfoot. This runaround means one has to get a Bigfoot no matter what. Nothing else is evidence. Everything was being sent to Slick's labs for analysis, so it seems certain that Slick was far ahead of the readouts than Titmus, Green and Dahinden.

In any case, Slick called-in Peter Byrne, whom he had known from his previous interest in the Yeti. Byrne had been a big game hunter turned to game preserver who had taken up the most protracted interest in trying to locate the original stimulus for the modern hoopla in the Himalayas. He was convinced the evidence for an American Yeti in the Pacific Northwest was much more than in the Himalayas, and he was preparing to get involved when exactly 4 years after the beginning of Bigfoot on October 6, 1962, Slick's plane crashed in Montana killing the Texas millionaire and PNE's financial backing.

With Slick's death PNE fizzled out, and the most credible field investigator, Byrne, went back to Nepal. Without Slick's financial support Bigfoot research essentially stopped. Kirk Johnson, Ed Patrick and Gerri Walsh fade away and become mere footnotes in early Bigfootery.

There then ensued what John Green called the 'Quiet Years,' in which the Bigfooters generally fell into discontact and forgot about Bluff Creek and Humboldt County. Bob Titmus even moved away to the backwoods of Kitimat, British Columbia. Except for a few plaster casts that members Green, Titmus and Dahinden kept for themselves, all the data PNE had accumulated was lost, supposedly destroyed by Slick's business associates. They had never been sympathetic to Slick's interest in Bigfoot; and the material, which had consisted of hair samples, feces samples, plaster casts, and photos, was simply in the way.

More than disinterest this act by his associates might reflect inside knowledge that Slick's PNE was not functioning in any positive away. In fact, Slick might have even expressed his frustration. Slick was intending a massive expedition *a la* some old African safari. But 3 years had not provided enough conclusive evidence to even give a geographic location to the expedition. Because of Bigfooters' analytical inabilities the Saskahaua— the traditional area of Sasquatch— was not even considered. The Bigfoot media blitz had caused northern California to take its rightful place.

Slick had cut checks for some men who sent very little remarkable back. Titmus cast nothing more than that odd hourglass print at Bluff Creek. Whether those huge "human-like" prints which Titmus claimed hunters had told him about for years were like the legitimate Ruby Creek Print we have no way of knowing. No casts had ever been taken of those; and Titmus' penchant to 'story tell' severally weakens any of his statements and subsequent input. Quite possibly this tendency on Titmus' part did not escape Slick or his associates, and this may have led to Slick's headquarters having a low opinion of whatever data had come in. It seems certain no lasting relationship with "Big Foot" was maintained by Johnson and the 2 other PNE members within Slick's organization.

These 'Quiet Years' ended in August 1967 when two different pairs of tracks reappeared on Blue Creek Mountain Road overlooking Bluff Creek. Not since August–October 1958 had two tracks been found together. Green and Dahinden returned. Green was stunned. The larger of the two prints he recognized as the 1958 Sandbar Prints (Hourglass Prints). "They were familiar to me— the same 15-inch print with a split in the ball of the foot that I had first seen 9 years before." "But the other print was not the Crew Print

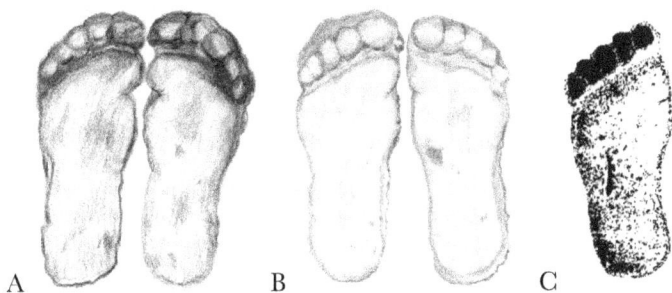

*A, Titmus casts from the Bluff Creek sandbar, 1958; B, from Blue Creek Mountain Road Prints, 1967. C. Bitmap of the smaller, 13 inch model.*

or anything like it. Rather, it was generally a smaller version of the Sandbar Print. It was 13 inches in length and also had that same curious pointless groove in the ball. It is called the Dahinden Print because it is preserved in a perfect photo which René Dahinden took.

This clinched it for both Bigfooters. The same Bigfoot which had caused so much excitement 9 years ago was still active in the Bluff Creek area and now it had a companion!

Bob Titmus, René Dahinden and John Green were the only PNE Bigfooters to reemerge from obscurity and start considering Bigfoot's reality again. They emerged, however, not as data gatherers but as hunters.

Although each one was far removed now from the hotbed of Humboldt County, none were ignorant that a new face in Bigfoot research had also appeared, alone in his pursuit of the hairy giant. His name was Roger Patterson. Roger Patterson was clearly the new boy on the block, but he wasn't the big kid, and the block which had captivated people's attention for a brief moment in 1958 was no more a stage. Interest in Bigfoot had died out and the audience had all but left. The block was still Bluff Creek. But thanks to the Blue Creek Mountain tracks of August 1967 the stage was once again Bluff Creek. It needed a new star. Bigfoot and Patterson were ready to debut.

# CHAPTER 4

# Bluff at Bluff Creek

In 1959, rodeo rider Roger Patterson, like millions of others, read that groundbreaking article by Ivan T. Sanderson in December's edition of *True Magazine*— "The Strange Story of America's Abominable Snowman"— and ever since that moment he was possessed with the bug to locate the elusive hairy giant. Patterson had a couple of advantages. One, his hometown was Yakima, Washington, in what could be viewed as the heart of Bigfoot country. Two, he was a horseman and woodsman and therefore he could get along in the wilderness. He was a small guy, about 5 foot, but he was sturdy and tough.

Patterson first made his presence known in 1966 when he self-published a paperback book, *Do Abominable Snowmen of America Really Exist?*— a delightfully sincere and privately printed compendium of newspaper articles and narrative that preempted even veteran John Green, who had been in the search for about 9 years and hadn't as yet written a book. This lovely bit of campfire Americana was illustrated with Patterson's own expert drawings of contemporary and classic "Bigfoot" incidents in tableau, making this the first book written exclusively on the subject of American "snowmen" a good primer.[22]

Patterson's title reflected the influence of Sanderson's works, to whom he devoted special thanks, and, moreover, the name Bigfoot was still largely obscure compared to the more popular Yeti. Patterson's book was also dependent on Sanderson's ideas that these "snowmen" were sub-humans. Patterson's expert sketches complied and made the Bigfoot just such a soul, complete with enlarged human feet, and in a number of sketches he flirted with the cone-headed image. America's Abominable Snowman was through Patterson's eyes an unusual hybrid.

Since Ivan T. Sanderson's article in 1959 nothing much had happened, causing Patterson to pen that ". . .one year went by— then two, three, four years— nothing on the giants of Humboldt County. Not even another magazine article refuting it as a hoax. Why surely," he continued, "I reasoned, that if it had been a prank or a hoax of some sort someone would have jumped on it."

In 1964, Patterson therefore tried to rekindle life into the hunt. He went to Bluff Creek for the first time to take up active investigation. This yielded, by direction of local Pat Graves, the finding and casting of one of the large footprints recently uncovered there on Leard Meadow Road. This proved to be yet another radically different foot from anything cast by the now-defunct PNE.

Leard Meadow Rd.
1964

The struggle to revive interest in the years afterward was even worse despite that December 1965 San Francisco *Chronicle* series "Animal-Men of the Northwest." During the 'Quiet Years' Patterson was just about the only man hunting Bigfoot in Humboldt County. . .until August 1967 when the Blue Creek Mountain tracks brought the disillusioned old Bigfooters back.

Inspired by the fresh news, Patterson also journeyed to Bluff Creek in October, ready to probe more deeply into the forest than they had. Green and Dahinden had only followed the tracks, measured them, took pictures and then left. But Patterson had something else in mind. His book had not panned-out as well as he had hoped,

---

[22] It should be clarified that some of Patterson's best tableaus were actually his own sketching copies of illustrations by Louis S. Glanzman and Mort Kuntsler in *True*'s Dec. 1959 and March 1960 articles respectively.

[ 75 ]

and he wound up unable to fund his own Northwest Research Organization. He now was trying his hand at making a documentary in hopes of garnering the hitherto elusive spondulics and sponsorship. Bluff Creek's recent events pointed to a Bigfoot near at hand. This is what he needed.

In a rather disturbing coincidence, Patterson now added to his firsts by indeed *filming* a Bigfoot on this very expedition. The date was October 20, 1967. With frequent sidekick Bob Gimlin, they were on horseback and approaching the creek bed when suddenly Patterson's horse reared from fright and fell over. He picked up his trusty 16 millimeter camera and started running while filming what had caused the horse to fright.

The film, mostly jumbled, captured some dark mass, tall and thickly built walking away at an angle. Out of the footage's mere 38 seconds[23] only a few seconds were clear and steady when Patterson stopped running and kept the lens on the hairy form. During this period it is plain to see that there was a large hairy Yeti-like gorilla walking away across the river bed. It was not frightened, but quickly walked to the other bank and into, presumably, the safety of the sequestering dank woods. (One must assume this because Patterson's camera ran out of film before the creature got there.) At one point, while fleeing from Patterson, the creature looked over its right shoulder at him. It was not out of fear but out of caution, keeping a trained eye on Patterson. When the creature turned, it was clear it was a female with large pendulous breasts. The 'Bigfoot' had an indistinct human face, was covered with hair, and its head was a bullet-head, as Sanderson popularized such a trait in his book, coming to a slanting cone. This is called a sagittal crest by zoologists and it is a characteristic of *male* gorillas.

Everything about the Bigfoot's appearance, coincidently, was like the popularized Yeti, but everything about its behavior, coincidently again, was identical to the Sasquatch that William Roe had described meeting in far-away British Columbia up Mica Mountain.[24] It would not be an exaggeration to say that this was almost an exact reenactment. The exception was that cone-head, that ever-so-obvious improv. from the Yeti that was horribly out-of-place on a female primate.

---

[23] Some say 39.2 or 39.7 seconds.

[24] Patterson also really didn't know Izzard and Stonor's works. They puzzled over how the Sherpas only saw male Yetis, never the female. If the Yeti is just a giant ape, as they speculate, then it would be probable that the female doesn't have a sagittal crest.

Patterson went on to indeed make his documentary of the exploit entitled, obviously, *Bigfoot* (1967), containing the electrifying footage. And one might add that every documentary since that moment has contained the short but sensational clip.

Patterson's film made the 1970s the decade of Bigfoot. For the first time the "animal-human" of legend really entered the vernacular of the American household as "Bigfoot" and permanently secured a place in American folklore. His film gave us the Yeti from the neck up, the breasts of Roe's Canadian Sasquatch, and the enlarged human feet of Bluff Creek. This became the single-o mug shot of Bigfoot. There were no more contenders to the image. Americans adopted this amalgam as their Bigfoot and allowed the Canadian Sasquatch to be the same thing. Roger Patterson would die of cancer in 1972, never fully seeing the effect that his "Patterson Film" would have, though certainly it had made his name go hand-in-hand with the entire pursuit of Bigfoot.

This is the pedigree of Bigfoot. This is the culmination of his evolution from the Yeti. It took about 12 years to give birth to him. It began innocently in the Himalayas with Tschernesky insisting the Yeti was a giant and it was abetted by American woodsmen who couldn't even distinguish the difference between the Yeti, Sasquatch and Bluff Creek prints. "The origin of such a creature is no problem either," boldly declared John Green in the conclusion of his groundbreaking first book. "Teeth and jawbones of something quite similar have been found in China. It lived 500,000 years ago, and has been described as being bigger than a gorilla but walking erect— a pretty good thumbnail description of what people have been seeing in North America all along. How it got here is also no problem. Man and a lot of animals are believed to have reached North America via a land bridge from Siberia. Gigantopithecus, as he has been called, could easily have tagged along."

Is the creature in the film truly the legendary Giant Ape? Even accepting it cannot possibly be the Yeti, can the amalgam in the film possibly be real?

A few scientists examined the film, and although they have been promoted by pro-Bigfooters to the world at large as remaining on the fence they were really on the fence about Bigfoot's existence, not about the film. John Napier, head primatologist at the Smithsonian, ultimately considered it one of the cleverest hoaxes. Though there are many grand quotes of him praising it, he wrote in his own book *Bigfoot* (1972): "There is little doubt that the scientific evidence taken collectively points to a hoax of some kind. The creature shown in the film does not stand up well to functional analysis. There are too many inconsistencies; yet no scientist to whom I have spoken and who has seen the film, has any direct evidence to prove that the episode was anything but what it was purported to be. My own comment quoted in the article for *Argosy* magazine (February 1968) was that there was nothing in the film which would prove conclusively that this was a hoax. In effect, what I meant was that I could not see the zipper; and I still can't."

Dr. Don Grieve, Reader in Biomechanics at the Royal Free Hospital School of Medicine in London, has appeared both impressed and in reactionary denial after examining the film. However, in studying the film closely, his problem was with trying to determine the speed at which the film was shot. If it was shot at 24 fps, then there was no doubt it was a man. Grieve felt there was little reason to suppose otherwise. Although Patterson had claimed he didn't know at what speed he shot the film, all knew he was intentionally making a documentary. Because 24 fps is best for TV viewing it is unlikely it was shot at any other speed.

However, Patterson's adamancy to commit himself to a clear statement as to what speed he shot the film has left the door open that it could have been shot at 16 or 18 fps, which at that speed means a man could not duplicate the gait of the "Bigfoot." Dr. Dmitri Donskoy preferred to assume this. He said that despite all the "diversity of human gaits, such a walk as demonstrated by the creature in the film is absolutely non typical of man." Dr. Grover Krantz practically devoted the second half of his life to repeated examinations of the film and was definitely a believer in its authenticity. Yet in countering the arguments from those who said its Bigfoot walked too much like a modern man not to be a hoax, he said: "If you have a large primate who has basically the bipedal locomotion he would have to have a human-like gait. This is exactly what I would expect"— thereby nullifying Donskoy's insistence that the figure in the film does not have a human-like gait.

"Expert contradictions" like this exist over every facet of the film, contributing to the erosion of confidence in its scientific analysis. But one thing none of them had was the computer technology with which to blow a frame of the film up and look at it. Each merely watched and estimated the stride of the Bigfoot to try and determine if a man could imitate it.

I was apparently the first one to get a copy of the famous Frame 352 (where the Bigfoot turns and looks at Patterson) and to blow it up using computer technology. What I discovered was both

*Sketch of Frame 352. Note the area of padding, plus the clear curve of the hunting gaiter used for the legs. The padding on the arm is the most noticeable, creasing severely with the backswing. This is hardly muscle definition; and indeed Roe said the Sasquatch had no corded muscle definition. Roe also made it clear that the females had no sagittal crest.*

amusing and slightly disheartening. It is a fake

With computer technology I was capable of finally getting a view of the face— that most crucial part which had so intrigued me— that which had never been described succinctly in all the historical cases. For this Bigfoot, its human lips are clearly visible; its nose is clearly European; the sagittal crest sits on its head like an upside down ice cream cone causing the hair to noticeably bulge; and the ears, so frequently mentioned in historical cases and represented in Indian artwork of the "wild man," are lost in the hair.

I could also get a view of everything else: the bulgy, tacky padding on the back of the arms, the hair covering the eyes, and the small hands remaining clenched, revealing no fingers. Also, if this was a real Sasquatch then it also seems to have the need to wear some kind of gaiters, for the upper leg clearly creases with the Bigfoot's stride, making a nice curving line over the outer thigh (plain to see even on the blow-up on T-shirts that a certain Bigfoot research organization sells). Its elbow also sticks out at 90 degrees—from a swatch of padding coming loose and bulging out.

The blow-up makes it plain that the suit was ingeniously done. The costume appears to be a padded body glove of some kind with the bulk of the padding on back. The armpit is clearly visible, only the shoulders and the back of the arm are heavily padded, causing the piece over the elbow to stick out at a 90 degree angle and the arms to look much longer than they really are. What appears to be a male pectoral can be seen, but it flexes too much with the movement of the arm not to be padding. In one frame it so creases just with arm swing the crevice is dark from shadow, visible even without blowing up the footage. The arm padding indicates that Patterson (presumably) knew he only needed to build up the back parts since the "Bigfoot" would be filmed walking away from him, as in the Roe encounter.

Right foot, Patterson Bigfoot

As impressive as computer technology is, and as impressive as the credentials of those who examined the film also are, neither is really needed. Independent of both is the knowledge of the actual footprint of the Sasquatch and the Yeti. Patterson's film was not just of something walking on celluloid. In making casts of the footprints, he and Gimlin left a tale as clear as any fingerprint. .These casts, unfortunately, were merely enlarged flat human footprints. They weren't what Roe described, what Dunn traced, what Shipton photographed, nor what any old newspaper had recorded.

More than anything else this film has created our modern image of Bigfoot and established northern California as its territory and Bluff Creek

Left foot, Patterson Bigfoot

as its shrine. Bigfoot and Sasquatch now became synonymous all over the world as a cone-headed Yeti. . .despite having the wrong feet a couple times over. Sadly, the Patterson Film became Bigfoot despite its massive and repeated inconsistencies with the historic data. It was the anchor; it was the glue that held together a pursuit that so far had turned up nothing concrete. Yet its Bigfoot's feet are its giveaway. It was inspired by the cu-

larged human feet of 1958 Bluff Creek, and its own footprints cast in plaster prove this.

Therein is the greatest controversy found in the 2002 obituary confession of Ray Wallace, Jerry Crew's boss in 1958. At no other time was the irony of the Bigfoot era more profoundly profiled. Within his obituary Ray Wallace's family reported that he was indeed the man who had made the footprints around Bluff Creek to hoax some of his lumbermen. (His brother Wilbur had been his loyal henchman, helping him on several occasions.) They confessed that everybody in their family had known what their father, noted for his practical jokes, had been up to, and now that he was dead they announced that "Bigfoot is dead." Ray's nephew, Dale Wallace, showed the wooden feet to the world. They were, in fact, those strange hourglass feet with that pointless groove in the ball, the very feet that Titmus and Green had cast in November 1958; the very same feet in August of 1962 which inspired the 1963 flap; the very same feet that also made the Blue Creek Mountain Prints in 1967 that brought Patterson there within a couple of months to film a Bigfoot that had never been there. The very seed that grew the Bigfoot tree upon which so many had nested was found to be false.

It would seem that this would be sufficient to finally bury the Bigfoot and Bluff Creek interlude and the remarkable folklore that sprouted the 9 foot tall cone-headed apeman with enlarged human feet. But many modern Bigfooters, that strange amalgam that tries to maintain every vignette of Old Bigfooter lore, refused to accept it. To confess that Ray Wallace's wooden feet were behind Bigfoot is not merely to isolate 1958 and eradicate it from Sasquatchery. Roger Patterson's Bigfoot clearly had feet that were inspired by these very famous "hourglass" prints, and he came to Bluff Creek to film something that obviously had *never* been there. To expose Ray Wallace's 1958 creation therefore is to completely undermine Patterson's Bigfoot and therewith Bigfootery's hallowed mascot and the entire phenomenon that spread over the Pacific Northwest thereafter as a result of his film.

The Wallace family confession raised a furor within Bigfootery for this very reason. Adamant Bigfooters even made footprint comparisons, placing pictures of the Crew Print cast next to pictures of the foot that Dale Wallace presented to the world. This accentuated how completely different the two were. This was indeed an accurate comparison. But it gently overlooked one real problem— both the Crew Print and the Wallace Print (the Sandbar 1958 Print) appeared simultaneously at Bluff Creek in August of 1958. Remember, Bill Chambers was shown both in order to verify Crew's initial story which was published on October 6, 1958.

Even if one wishes to accept the remote possibility that the Crew Print stands separate from the Brothers Wallace's shenanigans, it remains damning for Bluff Creek to admit that Wallace's hourglass foot appeared at the same time and continued for 9 years in the area without any Crew Print competition after that first August. Moreover, comparisons are deadly when properly done. Had modern Bigfooters done the same comparison between the Crew Print and the Ruby Creek Print they could not have hidden the fact that the Crew Print could not be Sasquatch anyway. The same methods they used to expose the complete physical disassociation between the Wallace and Crew prints would expose the complete irrelevance of Bluff Creek in the whole Sasquatch pursuit.

This is confirmed far more profoundly in John Green's old books, where text, photos and their captions bear witness to the truth that the Wallace foot *is the creator* of all the tracks he ever saw at Bluff Creek. In his 1968 *On the Track of the Sasquatch* Green even declares: "An unusual feature discernable in most tracks, in varying degrees, is a division right in the middle of what appears to be the ball of the foot" — an excellent description of the Wallace fake foot, a rather fatal declaration for Bluff Creek.

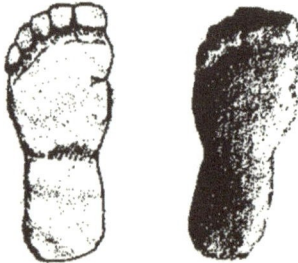

*The Wallace Foot left compared to a bitmap of "a typical print of the 15-inch track on Blue Creek Mountain road."*

In light of this, John Green's denial that Wallace could have had anything to do with the "Big Foot" of Bluff Creek is remarkable. Both at the time of the press furor and in his book's 2004 updated introduction he came out and unequivocally denounced how ludicrous the claim was to begin with, seeing how different the Wallace fake foot is in comparison to the Crew Print. Green so much as lamented, in what could be said to be a tone of frustration, that ", , , not one newspaper would so much as have a reporter talk to someone, namely me, who told them he had investigated the original incidents back in 1958, and had ample proof that Ray Wallace and the wooden feet could not have been responsible."

Ample proof, however, is simply not to be found. This was evident in Green's 2004 *The Best of Bigfoot/Sasquatch.* Instead of ample proof, a disturbing defense of Bluff Creek found its way into the subchapter "Big Foot Did Not Die." As far as Green could now admit the Wallace wooden feet did match the cast that Bob Titmus took on the Bluff Creek sandbar in November 1958. He argues, however, that they "are somewhat crudely carved, and presumably they were made in imitation of those casts." The assertion being made, of course, is that the strange hourglass print does belong to Bigfoot after all (despite its radical difference with the Crew Print and Ruby Creek Print) and the wooden feet Dale Wallace showed the world are merely wooden copies of an original Bigfoot cast in plaster and not the makers of the tracks to begin with.

The pictures used to support these bizarre arguments, however, are completely bogus. What is presented as the Wallace wooden foot and the Titmus cast of 1958 are obvious frauds. One must merely check Green's 1973 book *The Sasquatch File* to see Titmus pictured with his casts. The cast Green promotes as the Sandbar Print appears to be a cast of the Dahinden Print, 1967. The picture of the wooden foot Green promotes as the Wallace foot is one of many crude wooden copies that were carved by whittlers and sold as souvenirs because the hourglass tracks at Bluff Creek were so famous as "Big Foot."[25] There is a wood grain split in the heel of the foot that is a dead give-away it is not the Wallace original.

In light of how widespread are the photos of Dale Wallace with the wooden feet, held sole up for the news cameras, the claims committed over pages 14-15 of Green's 2004 book are profoundly disturbing and indicative of a complete lack of visual analysis. Green would never have engaged in chicanery; that must be made clear. It is more than likely that a member of modern Bigfootery fed these pictures of fakes to the aging Green. In doing so the Bigfooter did a great disservice to the man whom they uphold as the "First Man of Bigfoot Letters."

The pragmatic, hard-boiled portrait Sanderson had painted of Ray Wallace was worth thousands of words in favor of "Big Foot" in 1959, and when even back then it was rumored that Ray Wallace was behind the hoaxing the very fact that "Big Foot" almost

---

[25] Wallace also had a few varying pairs; he's even pictured with this one.

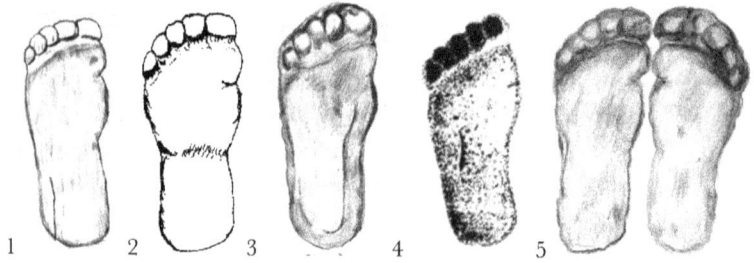

*1, the foot John Green says is the wooden Wallace foot; 2, actual Wallace Foot; 3, cast Green claims is the 1958 Titmus cast from the Bluff Creek sandbar. 4, the 1967 "Dahinden Print," the smaller 13 inch model. 5, actual sandbar casts from the Titmus Collection, 1958.*

destroyed his business made such an idea ludicrous. Such an argument was even used after Wallace's death to accentuate how preposterous it is that Wallace would have almost "cut his own throat."

The facts are that Ray and Wilbur were the sole source for this legend of loggers quitting. Always writing from afar, Sanderson was the perfect dupe and conduit for such a legend. Locally, the *Humboldt Times* preferred, with rare exceptions, to play-up how Bigfoot was rather benign. Betty Allen, the reporter-at-large who closely followed it all, declared in an October 7, 1958, edition that "if it is the work of an oversized human being, he is minding his own business and not hurting anyone." A number of articles stressed the harmless curious nature of Bigfoot. This characteristic was deduced by the fact Bigfoot's footprints actually showed something just walking past and around the logging equipment without touching it— the opposite of what the Brothers Wallace reported to Sanderson.

After Wallace's death, John Green also made a weak attempt to paint Wallace as anything but pragmatic and hardboiled. Some of his ludicrous old stories were dug up and used to portray him as having been an unstable man. But the truth of the matter is that Ray's letters and reports of sightings show a man more mocking Bigfootery than unstable. He continued on for decades, sending Green and others reports of his sightings and funny pictures of obvious fake Bigfeet. Only in this area did Ray overplay his hand, becoming something quite unbelievable. By the 1980s the vogue of Bigfoot was slowly moving on to giant apes in Grover Krantz's hands, and dermal ridges were becoming vogue on hoaxers' feet. There was something quite dated and early Bigfoot in Ray's continuing masquerade.

Patterson's film had represented museum old tracks anyway. It gave Bigfootery its new chapter: Ciné du Sasquatch. That piece of film became more important than any old track, and footprints became boring by comparison. Wallace was smart to bow out and frankly get lost in the fury of Bigfoot's very marketable decade. He wisely hung up those feet and there on the shelf they remained until his nephew Dale brought them out for the cameras in 2002.

Funny human feet, however, would remain the foundation, and Ray is the source for this. This and. . ."Later, Ray Wallace also stumbled upon a similar enormous mass of human shaped droppings," had written Ivan Sanderson in 1959. "He shoveled them into a gallon can and found that they occupied exactly the same volume as a single evacuation of a 1,200-lb horse!" Human feet and loads of, well, you know, would forever remain the foundation of Bigfoot thanks to Ray Wallace.

The "Quiet Years" actually coincide with Wallace moving from Bluff Creek to Toledo, Washington, where in 1964-65 the only man looking for Bigfoot, Roger Patterson, would visit him and hear of more hair-raising stories. It was through Wallace that Patterson was finally directed to the aged Fred Beck. At last Beck would be persuaded to tell his story of the "gorillas of Mount St. Helens." During a short trip in August (Wallace often hoaxed in August) 1967 to Bluff Creek, Wallace saw fit to lay down the last track of his now-famous hourglass print at Blue Creek Mountain.

Wallace's tracks and influence therefore cannot be downplayed. His "flap" of tracks came at crucial times, both inciting the development of PNE in 1958, whetting its appetite thereafter, inspiring the 1963 "flap," and in 1967 setting the stage for Patterson. In a very real sense then Wallace is indeed behind a large portion of Bigfootery, at least at Bluff Creek. . .and because of Patterson's capitalizing film the entire world. The posthumous confession of the lumberman's family was no lie. In 2002 Bigfoot died.

But in the interim, thanks to Patterson's film, the age of Bigfoot had come. But although the film prolonged the life of Sasquatch in the popular forum, it also maintained the fiction it was the Yeti, and in that way it did truly destroy the actual Sasquatch. In its place there were several beasts, creatures, or nightmarish forms. All of them were created in the image of those who wanted to see him in whatever form they fancied, all united only by some outlandish cone-head. They were silly, ridiculous; they were carnival and quite unbearable, and this, the 1970s, became the golden age of Bigfoot.

# CHAPTER 5

# Willow Creek Days

As radios were playing the tune 'Age of Aquarius' the new age of Bigfoot fever was striking. Perfectly situated at its fore were the old Bigfooters of the long defunct Pacific Northwest Expedition. And with them, of course, came their venal goal: slaughter a Sasquatch to reap prestige, glory, and always money. It is an interesting and a humorous irony that American fad at the time was to get away from the big cities and back to nature. To the days of man's innocency went scores of VW buses loaded with hippies anticipating the new age of brotherhood. Yet here in the rugged outdoors they had a very real chance of stumbling across woodsmen ready to shoot anything that moved for the exact opposite motive of the counter culture's arrival in the sticks.

This new age was one of capitalizing on Bigfoot. Traipsing over the woods in a dedicated attempt to find one (or to collect data) is a nice image, but this is not what Bigfooters did. Soon after filming "Bigfoot," Roger Patterson had set his course to income. In hopes of augmenting his "precarious living," as John Green put it, as an inventor and promoter, he was soon busy showing and shopping his movie at universities and connecting with any positive press he would. René Dahinden and John Green, too, saw the potential in the film. . .but as a source of income. Rather than viewing it as a key to starting an expedition, they agreed to split the costs to obtain from Patterson the rights to the Canadian lecture tour. Nothing seemed leftover for Bob Titmus. So he went back to the wilds of remote Kitimat. But in a mind-boggling paradox, no Bigfooter staked-out Bluff Creek at all. (No more than a couple of local hunters took pictures of the Patterson Bigfoot tracks).

This is one of the greatest paradoxes in the whole history of any pursuit. What in fact did Bigfooters believe? That is hard to say. There was neither method to their madness nor consistency to their method. All appeared to dogmatically believe the Patterson Film; that being the case Bigfooters had proof not only of Bigfoot's existence *but of its existence at Bluff Creek.* Yet none staked the area out. No camps keeping a constant vigil were erected. No scopes scanning the glades and creeks from mountain shoulders. No trailers housing round-the-clock observers and hunters ready to sortie out. The hunt for Nessie at Loch Ness inspired just such scientific endeavors, yielding as their result some interesting film and a number of sightings. But the Bigfooters forsook Bluff Creek for the whole of the Pacific Northwest. They used and reused Bluff Creek and Patterson's film for one thing: to anchor the entire popular phenomenon they were trying to cash in on.

Doubts, however, regarding the authenticity of the Patterson Film must certainly have initially been smoldering within Dahinden and Green. As the film's significance was becoming apparent, both wanted in on the lecture rights but only if it was legit. This may have prompted what was otherwise an inexcusably tardy investigation of the filming site. Alone of the two, Green made it to Bluff Creek in June 1968, hooking up first with Jim McClarin, a local Bigfoot enthusiast, to assist him at the creek bed. No tracks, of course, were visible anymore, as the winter floods had washed over the sandbar. But depressions where casts were taken, according to Green, could still be seen. Green, however, was not after evidence. He wanted to reenact the filming as one way of authenticating it. In doing so, he confirmed something that should have been considered alarming: Patterson had actually filmed the "Bigfoot" entirely from a crouching position. The effect would be to make the Bigfoot look bigger.

Green's reenactment was a rather complex method with a paradoxically simple goal of determining the Bigfoot's size and height—i.e. if too tall or too big it ruled out a man and indicated a Sasquatch. He had Jim McClarin re-walk the Bigfoot's path while he filmed from the "exact location." McClarin was 6.6 in his boots. After developing the film, Green traced a frame of the Bigfoot and a frame of McClarin when in the same approximate positions. Overlaying these indicated the Bigfoot was about 5 or 6 inches taller than McClarin, or about 7 feet.[26]

The Patterson Film, however, shows that in the frantic rush forward Patterson jumped over a small ridge, where he then stopped; given the soft sand his boots were probably several inches embedded into the sandbar. In a step-by-step forward passage of the Patterson Film frame-by-frame compared to the various positions of McClarin frame-by-frame in Green's film, it is entirely clear that Green held the camera too high. Green was also 1 foot taller than Patterson. What's worse is that McClarin is back further than the Bigfoot, which would make him look shorter when compared to the Bigfoot in the superimposition of the frames. The combination of discrepancies at that distance makes for noteworthy error in comparing their respective heights, possibly as much as several inches.

Drs. Grieve and Napier's studies were a little more impressive. The former's examination concluded that the Bigfoot was indeed only 6.5 feet tall. Napier summarized: "A stature of 6 ft. 5 in. is fine; there is no reason to exclude the Sasquatch on these grounds. *But* the footprints associated with this creature are totally at variance with its calculated height. The footprints are said to have been between 14 in. and 15 in. in length. On the basis of the coefficient . . . [stature= greatest foot length X 6.6]. . .this should equate with a stature of 7 ft. 8 in. -8 ft. 3 in. The space (the step) between one footprint and the next is given at 41 in. A creature 6 ft. 5 in. in height should have a step of 45 in., particularly, as it is seen in the film, when striding out; in fact in view of the exaggerated nature of the walk, the step might be expected to be somewhat longer than the normal, say 50 in. The conclusion is inevitable. The footprints must be fakes or the film is. Of course, both film and footprints could be faked but one thing is certain; they both cannot be truebill."

Basically, this was Napier's polite English way of saying both were fakes, for if the film was real why would somebody come along and cover up real Sasquatch tracks with fake prints? Napier's comments could also reflect a certain amount of inside knowledge about

---

[26] The logs had not shifted height. It would have been better to stand a ruler on top of the log and then superimpose that picture on the frame where the Bigfoot is right behind the fallen log. With the height of the log known, it would have been a closer measurement.

the immediate and suspicious actions of Patterson and Gimlin after they shot the film. These same suspicions have smoldered for decades even amongst a very strong pro-Bigfoot contingent.

It came to a head in Greg Long's much-maligned 2004 book *The Making of Bigfoot*. The book showcased the confession of a man named Bob Heironimus. Scuttlebutt going about in the "culture" of Bigfoot in the Pacific Northwest had long pegged him as the man inside the Patterson Bigfoot suit. Heironimus finally confessed in Long's book and detailed what went on that day in 1967.

In his recollection of the events, Heironimus states that the film was shot earlier than the 20th of October. Also, he said that Patterson gave the film to him with the instructions to take it and mail it from Eureka to Yakima, Washington, before going home to Yakima himself. Patterson, according to Heironimus, said: "We have to go back and make them [the tracks]. We'll either do it today, or tomorrow, and we're out of here. . ."

Although Heironimus' confession is challenged by some (and is indeed shaky in areas), this last statement has a solid and resonating ring of truth to it for several reasons. One, it seems the only explanation to account for what Grieve and Napier turned up regarding the fact that the footprints weren't made by the Bigfoot in the film. Another reason this part of his confession should be regarded as true is the known chain of events. The most accurate account actually comes from Don Hunter and René Dahinden in their joint book together, *Sasquatch* (1973). Dahinden was in San Francisco when the news broke and he immediately headed north. "René arrived at Willow Creek the day after the filming," Hunter writes, when he came across and surprised Patterson and Gimlin in town (neither had known he was in San Francisco). Both quickly claimed they had just gotten out of Bluff Creek before a deluge caused the creek to rise quickly. They had had a "hair-raising trip out, their truck on several occasions coming frighteningly close to sliding off the narrow muddy road on Onion Mountain." As a result, Dahinden was sufficiently discouraged from heading to and checking out the site. According to Patterson, the reason they were still in Willow Creek was because they had already mailed the film from Eureka and had come back merely to cover some footprints to preserve them (and cast one).

Thus the sequence is: Patterson/Gimlin supposedly shot the film, drove to Eureka (an hour drive of about 50 miles along a narrow mountainous road), mailed it to his brother-in-law, went back to Bluff Creek, up again and over Onion Mountain, to cover some tracks for future casting, barely got out during a rain storm, met Dahinden, then all three went to Yakima.

As suspicious as this chain of events is, it gets worse. While in Eureka, Patterson had alerted the newswires. This is a remarkable thing to do after mailing undeveloped film, given that Dahinden and Hunter write: "The film was being processed and Patterson, as he told René, didn't know whether he had anything that would stand up." Hunter continues: "Next day, the group viewed the film at the Yakima home of Patterson's brother-in-law." Overlooking the illogic of mailing the film, the unnatural speed with which it made it to Washington and got developed, there is the question of alerting the world press. No one would do this unless he was sure he had something worth showing. Critics' who claim that Patterson had already shot the film and knew it could work seem justified in their suspicions. There are many reasons to believe he merely was staging footprints on October 20.

The only logical part in this is that Patterson staged his film at Bluff Creek. Bigfoot was really only associated with this area, and the local *Humboldt Times* had shown a ready interest to report most anything Bigfoot. Bluff Creek was a quick "in" to the newswires and from there to national news, especially after the Blue Creek Mountain Prints of August (thanks to Ray Wallace).

The one truth we do know is that there was no immediate investigation of the film site. Dahinden was thwarted merely by the claim of possible mud slides. Green later wrote that he spent too much money trying to get a scientist to tag along with him, and thus he ended up too short of funds to go to California. Both Green and Dahinden therefore had to rely on Bob Titmus. What an ironic stroke, for he was, to put it politely, the most "honesty-challenged" of all the Bigfooters. . .and this would cause serious problems for the film later on.

Titmus drove from as far as Kitimat (about 1,500 miles) to get to his old stomping ground of Bluff Creek 9 days after the supposed filming. Titmus reported that he could easily follow Patterson, Gimlin and the Bigfoot's tracks. It deserves to be quoted from Titmus' own letter to John Green, which he reproduced in *On the Track of the Sasquatch* (1968). Titmus' letter contains some very interesting information. In it, Titmus writes that the "tracks traversed a little more than 300 feet of rather high sand, silt and gravel bar which had a light scattering of trees growing on it, no underbrush whatever but a considerable amount of drift debris here and there. The tracks then crossed Bluff Creek and an old logging road and continued up a steep mountainside. . ."

About 80 or 90 yards up into the deep carpet of ferns and under-brush Titmus found "very plain evidence where Bigfoot had sat down for some time among the ferns." Titmus speculated that the

#1 Titmus cast
of Bigfoot

Bigfoot was watching Patterson and Gimlin while they were down on the sandbar casting the tracks. "The distance would have been approximately 125-150 yards. His position was shadowed and well screened from observation from below." Titmus also claimed that the tracks "continued on up the moun-tain" although he didn't bother to follow them. He also admitted that he "spent little time trying to back-track Bigfoot from where his tracks appeared on the sandbar since it was soon obvious that he did not come up the creek but most probably came down the mountain, up the hard road a ways and crossed the creek onto the sandbar," although he does not clarify how he knew that.

Of his study of the footprints that still remained (he said that Pat-terson and Gimlin had covered a couple with bark so that they remained in good condition), even though 9 days had elapsed he was able to assert that he could detect that "Most of the tracks showed a great deal of foot movement" although also "some showed a little, and a few indicated almost no movement whatever." Despite spending hours examining the site (besides casting 10 consecutive prints) he did not count all of them or measure the stride between them or even bother to look for hairs where the Bigfoot supposedly sat down. [27]

#2 Titmus cast
of Bigfoot

On the face of it, Titmus' claims add dynamic support to the Patterson Film. In substance, howev-er, he had nothing to show for it except 10 plaster casts, a rather improbable story and, worse, he had inadvertently confirmed that Bigfoot's tracks really only existed on the sandbar.

This would become a sore point years later when Dahinden was firmly in place as the majority owner of the film. He, in fact, be-came very sensitive about the fact he had only film and no corrobo-

---

[27] Titmus mentions no storm damage in his letter to Green. "I spent hours that day examining the tracks, which, for the most part, were still in very good condition, considering that they were 9 or 10 days old." Could Patterson and Gimlin have been trying to keep Dahinden from the site with stories of mud slides?

rating evidence from the site. When Greg Long had asked him about Titmus' findings at the Bluff Creek site he crudely retorted: "Titmus vas a storyteller. I don't believe a fucking vord of it. You have to know the location." Long recorded the "cantankerous" Dahinden saying with disgust: "Titmus fucked up! He never counted the tracks. He never photographed them. He didn't even have a measuring tape along to measure the step length."[28]

Dahinden was hardly trying to undercut the authenticity of the Patterson Film. This was merely one vent in decades of venting that he had nothing but the film and thus, he asserted, the film must be and should be judged on its own. Even Patterson's reputation as a "crook" didn't matter anymore, although it was clear Dahinden remained sensitive about it. "I don't care if he's out there fornicating with his sheep. Look at the film!"

#3 Titmus cast
of Bigfoot

Not surprisingly Dahinden's belief in the film had grown with his ownership stake in it. Despite his initial cautious attitude about it (as expressed in *Sasquatch*, 1973), Dahinden saw how the film was becoming increasingly popular. Even before he was an owner, he had gone to the lengths of taking the film far and wide (even to Europe) to get those much quoted but ambivalent conclusion of scientists. Then after Patterson's death he finally persuaded Gimlin to sell his share for a pittance and eventually through one means or another— although they all come down to one word: lawsuit— he obtained 51 percent share in the film.

. . .The fly in the buttermilk, however, is that by the time the film became so popular Dahinden had nothing. He had only faith in the film, and this put him in an unwieldy position to defend it. He had too long insulted Green's methods as "shallow" and degraded Titmus as inept to bring into play any of their supportive evidence for Bigfoot's behavior *off camera*.[29] But as unreliable as Titmus might have been, his casts were not. These clear imprints prove Patterson's claim of the deluge and the creek bed rising (and thus sweeping over the low sandbar) is bogus. In undercutting Titmus' report of the Bluff Creek conditions, Dahinden may have hoped that this insulated what he thought was sure-fire film from

---

[28] Yet in his joint book with Don Hunter *Sasquatch* (1973), it is asserted that the Bigfoot's stride was 40-42 inchs.

[29] Green's investigative approach was indeed cumbersome and suspect. Grieve's examina tion was only over his head, and Napier clearly thought there was fakery involved.

any extraneous challenge, but all the evidence together only proves that Patterson lied and Bigfoot only exists for 38 seconds on the sandbar, and this should be regarded as more suspicious than anything.

The set of circumstance that followed, as even Dahinden accepted them in 1973, are equally strange and suspicious. And his part in the aftermath, carefully positioned as Patterson's wise counselor, shows how money was foremost in the rodeo rider's mind. Patterson was so hyped up he wanted to go to New York and "stun the world of science, at the same time building dreams of making a vast fortune." Dahinden declared: "With him it had to be a million bucks or nothing." Dahinden cautioned against this. "If you take this to New York, they will laugh at you. You'll be just another guy with a monster movie."

The film is far from a cheap monster movie, as Dahinden's later obsession to own it proves. He may have stopped Patterson from going to New York, but Dahinden quickly seized the initiative to go overseas to authenticate it first. Altogether the cumulative effect is not good for Dahinden or Bigfootery's memory, for in his desire to capitalize on the film he reveals to what extent he would not question the *circumstances* of the filming.

Dahinden's venting in later years mutely declares how much he did not regard Patterson's own words as trustworthy. Surely, Patterson had measured the step length, did he not? He was specifically heading "deep into the bush" on this expedition (though this was actually very close to a road) to track one; he must have had a tape measurer or used some other method. It is only logical he should be prepared. But he didn't. He didn't even film them in a manner that would help judge the stride. By 1971, when Ron Olson was the centerpiece of *Bigfoot: Man or Beast?*— those halcyon days of Bigfootery— Roger Patterson was still alive and even interviewed therein. Had there been concrete measurements, John Green, René Dahinden or Ron Olson could easily have obtained them. Whatever Patterson told them, if anything, was certainly not believed by Dahinden else 30 years later he still would not need to castigate Titmus for not measuring the stride.

It goes on and on. There is no point in beating a dead horse anymore.

However, Patterson and the circumstances of the filming have become so sensitive an issue because the film continues to be challenged more and more even by a very "pro-Bigfoot" faction in the world of those who follow crypto-zoology. Pro and Con are equally amazed at what little investigation Bigfooters did, and this more than anything undermines the value of the data they built up, which constitutes some of the most hallowed pillars that continue to uphold the entire legend.

Furthermore, challenging the legend uncovers a disturbing Bigfoot culture. There is no hiding the fact that there developed a close knit group of "believers" who were out to protect the Bigfoot of 1970s' legend and with this anyone who contributed significantly to that image. Belief in the myth takes the place of belief in the substance that spawned it. To many this becomes like a religion, and every vignette, every popular account that becomes accepted must be endorsed. To deny one is to be a heretic. It's like saying I believe in Jesus but I don't believe the water-into-wine episode. You are a heretic.

It is this element, perhaps more avaricious than religious, that protected Bigfoot and Patterson's film from critical examination. Dahinden bluntly denied Long access to it, the retired milkman and slug tote declaring Long unqualified. Often there are elements in the Press that seek a cash cow. In this case Bigfoot provided a carnival version to take out occasionally. Coverage was seldom quite in depth or critical of the film. Early Bigfoot documentaries, moreover, were also made by very pro-Bigfoot contingents— Patterson's documentary, of course, which he showed on tour and narrated to live audiences; *Bigfoot: Man or Beast?* surrounded Ron Olson's group in 1971; and the granddaddy *Mysterious Monsters* with Peter Graves in 1975 made no bones about the Patterson footage being genuine.

Now that it seems both appropriate and more profitable ratings-wise for the media to expose much of the myths an earlier generation endorsed, denied access to critically examining the film has proven the core owners are an obstacle, and the caution and selectivity with which its imagery is disseminated continues to excite the belief that it is known to be a pure forgery even by those who control it.

Even though not subject to blow-ups back in the 1970s, the film's weak points are clearly visible even to a casual observer who can stop, slow and fast forward a copy on DVD. Actions often outright contradict the narration illustrating the viewer to something that is

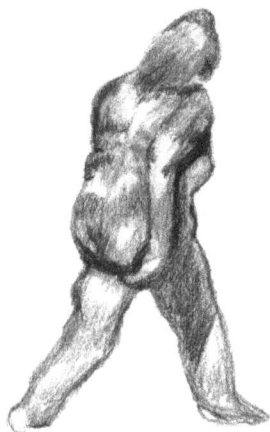

not there. The arms are simply not longer than human arms. They are the typical length of a tall man. Measuring from the armpit to the wrist makes this clear. The arm swing brings the hand down near the knee only because the "Bigfoot" is hunched over, lowering the arms in the process. This tendency to hunch seems more pronounced as the Bigfoot comes to the place where it will glare at Patterson. But several steps before that the Bigfoot is standing more posture upright, and it is obvious on the backswing of the arm that the hand is right below the butt just as on any human.

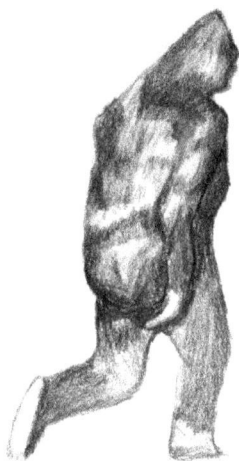

Greg Long's much maligned book *The Making of Bigfoot* was essentially an exposé of the Patterson Film and the culture surrounding it. Long's book, however, was maligned for basically being viewed as a long character attack on Patterson, which is not the intent of this author. I am not interested in Patterson's character, whether he was regarded as a "crook," "conman," "welsher," or "chiseler." His actions after-the-fact speak volumes. Like the other Bigfooters he never bothered to stake-out Bluff Creek. Rather, he followed the popular frenzy over the whole Pacific Northwest.

This frenzy was one of big-busted Sasquatches. As indelicate as that may seem to the reader, it is nothing compared to its euphemistic appellation of "Boobfoot." Film has the potential like no other medium to inspire the ambitions of the opportunist, copycat and nut.

The first major story broke on July 12, 1969. A gray-haired Bigfoot with "large breasts" watched Charles Jackson and his son Kevin burning rabbit entrails in their yard at Oroville, California. Because it was reported in the local paper, it set off a very clear chain-reaction. None other than Roger Patterson came by to investigate on behalf of his own Northwest Research Organization, in dutiful obedience to his sterling motto, in bastardized Old English, "He who seeketh long enough and hard enough will find the truth, whatever that truth may be." An elderly man named Homer Stickley then reported he had seen a female on a number of occasions. Homer saw one standing by a tree. Then he saw one swing 'them' across a field. He even impersonated her gait and hunched stature for a picture. Roger Patterson stayed at Homer's and set up his research headquarters there. Together they advertised for sightings reports in the local newspaper.

"Encounters in retrospect" soon surfaced. In mid June of 1968, near Clipper, Washington, a logger named Frank Lawrence Jr. claimed he had seen an entire family of nude black-haired Bigfeet. The woman was easily discerned by her large breasts. He ran off. In December 1968, at Clackamas River, Oregon, Glenn Thomas studied two female Bigfeet while they slept and ate together. One of the biggest— no pun intended— was on the night of July 26, 1969 (soon after the Oroville news coverage), near Hoquiam, Washington, when deputy Sheriff Verlin Herrington saw a big mama on the side of the road about 8 feet tall and with large breasts. Just from training his spotlight he could see that they were covered with hair but that her nipples were not. Herrington was not rehired by the Sheriff's Department when his year's contract was up. North of Orleans, California, summer of 1969, a woman claimed she frequently left apples and fruit out for a male, female and a young Sasquatch, who would come and eat it. In September 1969: Kananasis Lake (Ribbon Creek) near Banff, Alberta, a female Sasquatch squatted within 100 yards of three prospectors. She watched them and then got up, chattered her teeth, moved arms up and down, and then left. Her breast were described as long and droopy, her head "sort of pointed" and her hair "brownish." She was about 7 feet tall.

Curiously, no matter where Patterson went he found similar footprints. It is really the most interesting phenomenon that certain styles of footprints were exclusive to certain Bigfooters. In his first excursion into Humboldt County he conveniently cast the Leard Meadow Road Print which looks conspicuously like an enlarged version of his own foot. Then in visiting Fred Beck he finds and casts another very human-like 18 inch print at Erion Ranch, near Woodland, Washington. From the description of the "Sasquatch" by the ranch owner, Charlie Erion, Patterson declared it to be the track of the "hairy apes," borrowing the handle from the Mount St. Helens legend and old Beck with whose story he was enchanted. . . though he forgot the "apes" had 4 toes and not 5. After contact with Ray Wallace (who had set him on his course to Beck), Patterson's Bigfoot develops suspiciously similar toe arrangement to Wallace's famous "hour glass prints" at Bluff Creek. [30]

---

[30] They are "completely human except the toes are relatively small," Sanderson wrote of Bluff Creek's footprints in his pivotal 1959 article. We know Patterson devoured the article, but it seems unlikely that he knew what the hourglass prints looked like until he met Wallace in 1969. Who also had casts of them.

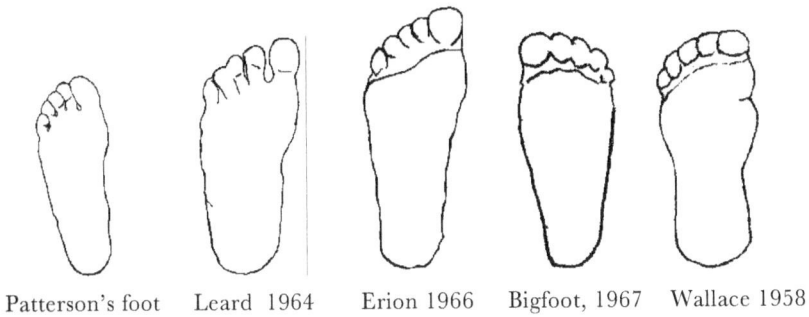

Patterson's foot    Leard 1964    Erion 1966    Bigfoot, 1967    Wallace 1958

In a strange way the evolution of Patterson's footprints mirror the old and become a prelude for what was to come after his film electrified the media. Big foot prints blazed a trail from Bluff Creek through the entire Pacific Northwest. They came in patterns— enlarged human feet and hourglass prints first (1958-1967); eclecticism (1970s); strange feet with dermal ridges (1980s); and today anything goes.

Opposite Page—

A Rogue's Gallery. All illustrations are drawn by the author straight from the actual casts or photos of them. These represent casts that have been accepted as genuine by all major Bigfooters. **A.** The Jerry Crew print, Bluff Creek, California, August to October 1958. **B.** Ray Wallace's "hourglass" Bluff Creek foot, August 1958 to August 1967. **C.** A print found at Hyampom, California, near Willow Creek, back in 1963 by Bob Titmus. **D.** The clearest print taken by Roger Patterson of the "Bigfoot" that he filmed on October 20, 1967. **E.** The other foot of the same Bigfoot cast by Roger Patterson. These are to scale. Many excuses have been offered and accepted to account why one foot is smaller, but none have been believable. **F, G, & H.** Bob Titmus took 10 consecutive casts 9 days after the alleged Bigfoot filming at Bluff Creek. These are three of them. **I,** the 1967 Dahinden Print, Blue Creek Mountain. **J, K, L & M,** foot casts (some by Titmus) over the Pacific Northwest; **L** I dub the "Weirdo Print." **M** looks close to the fakes made by Ray Pickens of Chehalis, Washington, 1975. **N.** The Terrace, British Columbia, Print of 1976. John Green considered this the clearest print in British Columbia. It looks like a bathtub sticker inspired it. **O.** The infamous "Cripple Foot," left and right feet. See Chapters 11 & 12. **P.** is the Indiana Print which was later admitted to be a hoax. **Q.** The cute Wrinkle Foot, left and right foot, 1984. **R.** A Blue Mountain Print found in 1982 by Paul Freeman. **S.** A Blue Mountain Print, 1986. **T.** The Tollgate, Oregon, Print, cast in 1986. **U, V & W** "Dermals" from the Blue Mountains in the 1980s.

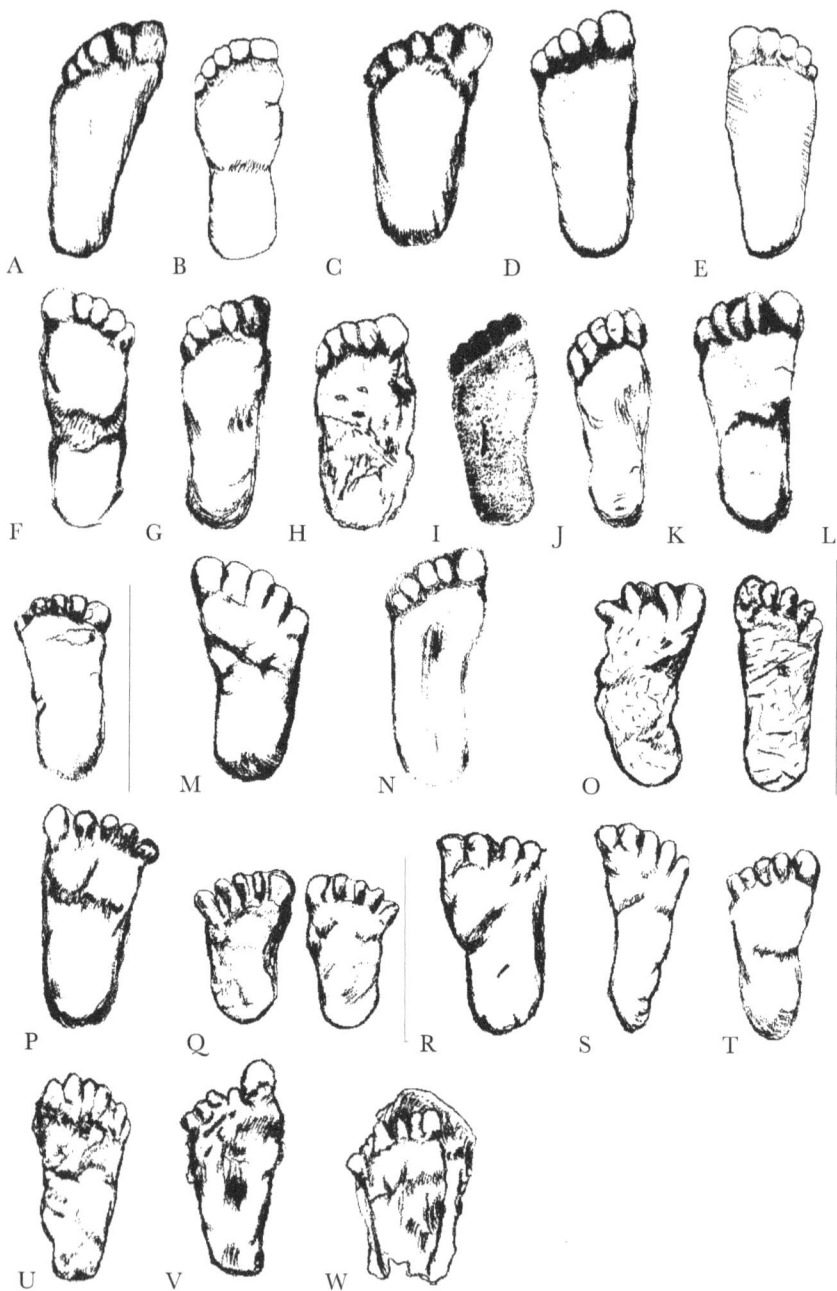

Willow Creek, California, made famous by Patterson's film, became a literal and figurative doorway to the Bigfoot phenomenon and the psychedelic cone-headed beast that popular culture was painting. There is no certain place where you can go to see a Bigfoot. True adventure and mystery has no guidebooks, of course. But beyond Willow Creek you are undeniably in "true" Bigfoot country. Any sighting report made beyond this point northward to Washington State carries more weight simply because of its location. Among all the false impressions that were created in the 1970s there is none more appropriate than this one.

This small California hamlet is located south of Weitchpeg and the Bluff Creek/Klamath Road area where Bigfoot officially began in 1958. This charming little place at the junction of Highways 299 and 96 advertised itself officially as the "Gateway to Bigfoot Country" and, by mutual consent, the citizenry even placed within the town an 8-foot tall wood statue of a Bigfoot carved by Jim McClarin. It's a dopey looking ape that would be at home in a Flintstones comic short, perhaps Barney Rubble's new pet. There is Bigfoot memorabilia, pictures, key chains, bumper stickers, footprints, pewter figures, imprints, drawings, mugs, a museum of plaster prints and other memorabilia, and you name it. Bigfoot is big!

For the very reasons it became the literal gateway to Bigfoot Country, Willow Creek became a figurative gateway to this era of merchandising the many faces of Bigfoot. It began here with Patterson's electrifying film and extended to the predictable world of copycat, opportunist and nut.

Oddball encounters with Bigfoot happened all over the United States, including Pennsylvania and New Jersey. There was a Bigfoot in Arkansas now (which inspired the 1970s cheese *The Legend of Boggy Creek*). There was the "skunk ape" of Florida. There were Bigfeet in Michigan, Maryland, Maine and Vermont. Kentucky had one, so did Ohio, Indiana and Illinois! The Bigfeet ranged in size from 8 to 10 feet. Some were white, others were brown, still others were black.

Just like "Boobfoot," many stories came out in retro. My favorite is Doc Priestley's account, supposedly from 1960, contained in Janet and Colin Bord's 1982 *Bigfoot Casebook*. It happened in West Virginia. He was driving near Marlington following a bus that had some friends in it. Suddenly his car sputtered and died. Then, suddenly, Priestly saw on the roadside a "monster with long hair pointing straight up toward the sky." Priestly confessed he was so transfixed on this creature that he lost track of time (and in my opinion quite a bit more). Finally the bus backed up to see what happened to him. That seemed to frighten the beast. Its hair dropped. Then it rushed off. Priestly didn't tell anybody what he saw. He just followed the bus again. Miles down the road the car began to act up again. Sparks were flying from under the hood. "And sure enough, there beside the road stood the monster again." Backing up a second time, the bus returned. The monster fled again. The Bords add: "The apparent ability of some Bigfeet to affect the working of electromagnetic devices may be linked to the question of whether they are physical or nonphysical creatures. . ."

Southern comfort happened to another man in Virginia in 1960. While in the woods near Davis, he and his friend went out to collect firewood. He then felt somebody poking him in the ribs. According to *The Bigfoot Casebook*, "he naturally thought it was one of his friends— until he turned and found himself face to face with a 'horrible monster.' 'It had two huge eyes that shone like big balls of fire and we had no light at all. It stood every bit of eight feet tall and had shaggy long hair all over its body.' "

Perhaps the worst story to become sacrosanct in the 1970s is that of Albert Ostman. It is perhaps the worst because it has become one of the most famous stories in the dossier of popular Bigfoot legend. Ostman had gotten into the game early. He had heard about the incidents at Bluff Creek and on October 22, 1958, wrote to the *Humboldt Times*. "I claim to be the only living man that seen a Sasquatch." What he wanted to know in particular was what the footprint looked like at Bluff Creek because he knew what the real footprint looked like. Ostman would later elaborate in his own account which John Green would publish in the Agassiz-Harrison *Advance* and then feature in his 1968 *On the Track of the Sasquatch* via which it came forward into the 1970s.

The dead give-away about Ostman's story is that he makes the Sasquatches subhumans after 1950s' examples of what cavemen were thought to have looked like. The other big difference is that they reflect White Man's fancy of being 8 feet tall and not Indian regard as 6.5 feet tall. Another one of Ostman's account's weaknesses is that it reflects a vignette in J.W. Burn's article in the popular men's magazine *Wide World* in January 1940, the kidnapping of Indian maiden Serephine Long.

Ostman claimed the incident happened in 1924. After a very improbable kidnapping in his sleeping bag, he describes them. The mother is notably ludicrous. "The women's hair was a bit longer on their heads and the hair on their backs had an upward turn

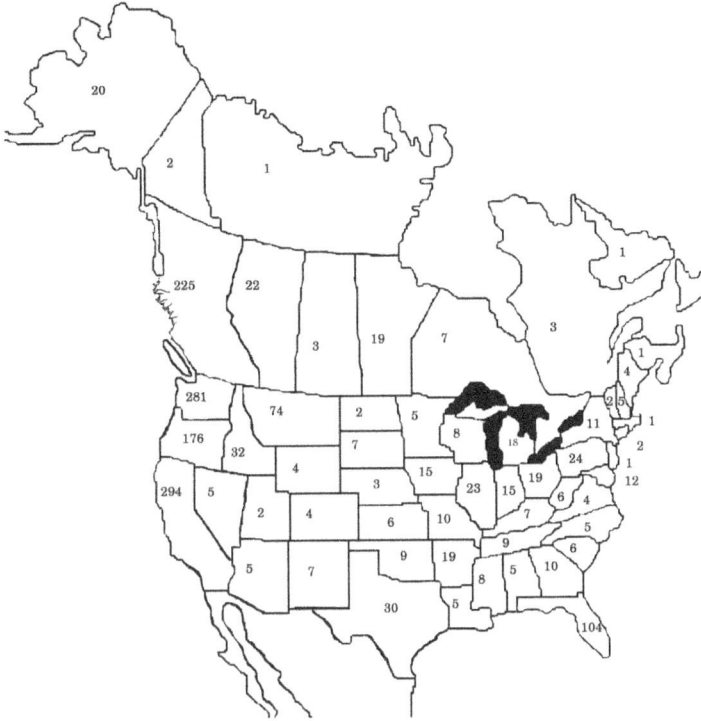

*At the height of Bigfoot mania in the 1970s, maps such as the one above reinforced Bigfoot's existence by showing how many "verified" reports of him had recently been made throughout North America.*

like some women have— they call them bangs, among women's hair-do's. Nowadays the old lady would have been anything between 40-70 years old. She was over 7 feet tall. She would be about 500-600 pounds. . .She had very wide hips, and a goose-like walk. She was not built for beauty or speed. Some of those lovable brassieres and uplifts would have been a great improvement on her looks and her figure."

Well, it goes on and on. The "old man" (father Sasquatch) liked to sit at the mouth of the valley hideaway, and when Ostman tried to stomp out the old Sasquatch put his arms up and grunted something that sounded like "soka, soka." Another elucidating word was "ook." Ostman didn't want to shoot him, so naturally he bided his time for several days and then finally made good his escape.

Having observed their lifestyles for those 6 days, Ostman talks about their behavior, which includes being fascinated by him making coffee; how the "old man" ate his chaw and licked the box with his tongue; and then other little literary snapshots of their daily and quaint family unit: the old man's discrete penis and huge big toe, which is all he needed to walk up a mountain.

Blunt but visual seems appropriate here. Though not wishing to offend I must nevertheless confess I am stuck without a better Latin phrase than *scattaurus* unless I tread into the world of colloquial English.

Yet although we may dismiss *Homo ostmanus* and White Man's crow about the odd and Cro-Magnon, there is something of compelling authenticity in Indian accounts of "Sasquatch men" we cannot dismiss. Laugh as we may about our own psychedelic incarnations, there is something about the old Indian accounts that is begging to be heard and taken seriously. It is about humans and a primate. It is about a pointless embellishment of 2 tribes constantly at war. It nags at one repeatedly.

Indian artwork bears silent but profound support to all what the Indians told J.W. Burns. For instance, the Kwakiutl Indians call the "wild man of the woods" *bukwas*. Sometimes he is portrayed as some fierce quasi-ape, baring teeth and snarling fiercely. But examples exist of the *bukwas* mask with a human face and workable jaw like a nutcracker, through which when wearing it the Indian shaman gave his advice. The face of one mask in the National Museum of the American Indian in Washington DC is one such example. It is human but still possessing the typical identifying features of the *bukwas*, such as beady, sunken eyes. He has high cheekbones, a sallow, dirty face, large protruding lips, a gaunt, gross and tormented appearance. Thin mustache and beard speak of sparse and unkempt hair growth rather than manicure. *Bukwas* in this case is indeed human, but he is no Indian and he is no White Man. The mask was thought to be carved in the late 19th century by the Salish Indians, the very same people of which those branches that live in the Saskahaua district of British Columbia call the resident wild man Saskahaua George.

While there is a slightly negroid appearance to the face, the sallow color and dirtiness strikes one more as a caveman, wild and primitive. There is nothing ape about it. Just as there is nothing ape about the actual Sasquatch men in Indian legend. They were hairy all over, but they were big men who spoke the Douglas dialect and resisted civilization.

In J.W. Burns' foundational article in *McLean's Magazine* ("Introducing B.C.'s Hairy Giants") he notes how Charley Victor regarded the negroid Busquatch woman to have psychic and super

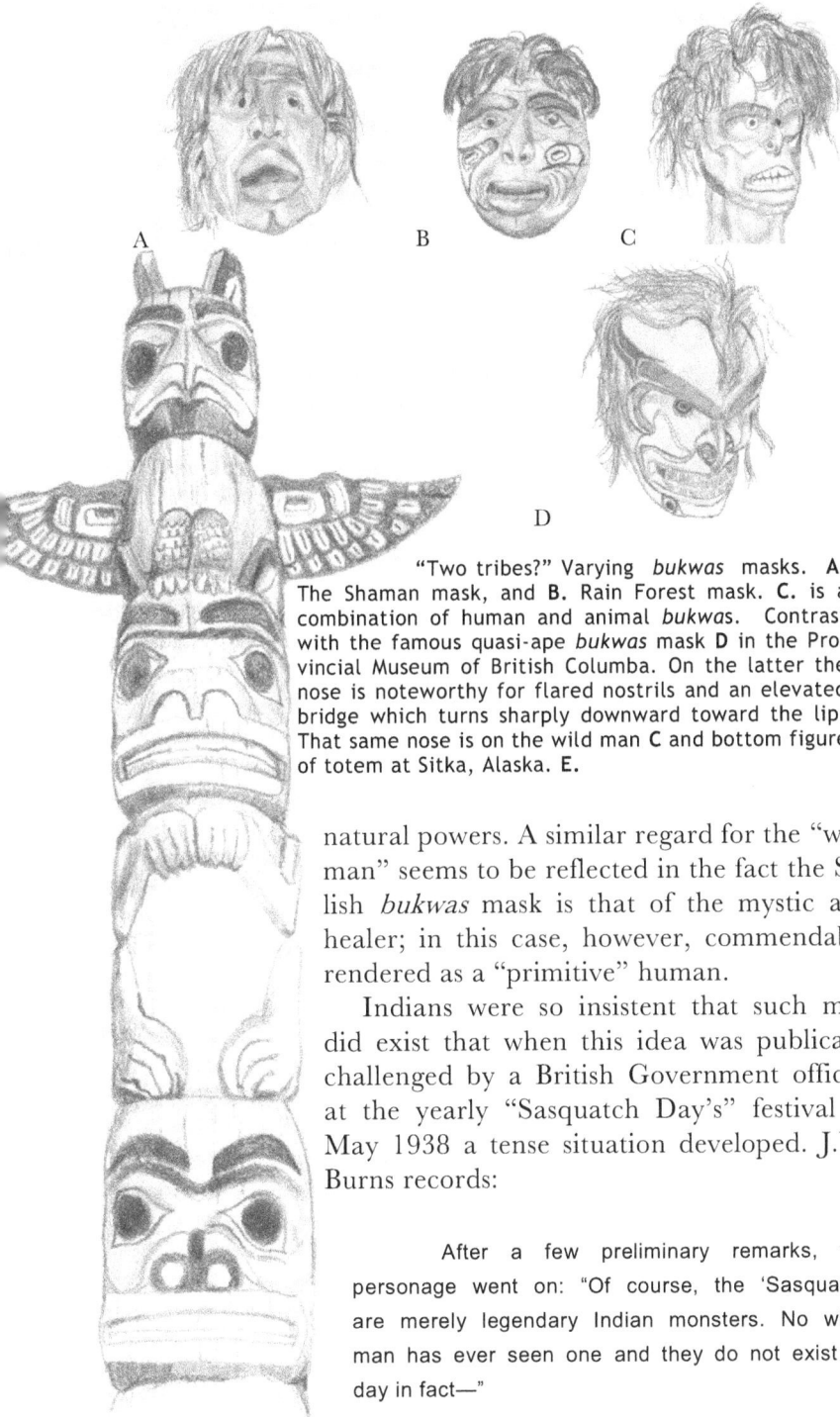

"Two tribes?" Varying *bukwas* masks. **A.** The Shaman mask, and **B.** Rain Forest mask. **C.** is a combination of human and animal *bukwas*. Contrast with the famous quasi-ape *bukwas* mask **D** in the Provincial Museum of British Columba. On the latter the nose is noteworthy for flared nostrils and an elevated bridge which turns sharply downward toward the lip. That same nose is on the wild man **C** and bottom figure of totem at Sitka, Alaska. **E.**

natural powers. A similar regard for the "wild man" seems to be reflected in the fact the Salish *bukwas* mask is that of the mystic and healer; in this case, however, commendably rendered as a "primitive" human.

Indians were so insistent that such men did exist that when this idea was publically challenged by a British Government official at the yearly "Sasquatch Day's" festival in May 1938 a tense situation developed. J.W. Burns records:

> After a few preliminary remarks, this personage went on: "Of course, the 'Sasquatch' are merely legendary Indian monsters. No white man has ever seen one and they do not exist today in fact—"

Thereupon his voice was drowned by a great rustling of buckskin garments and the tinkling of ornamental bells as, in response to an indignant gesture from old Chief Flying Eagle, more than two thousand Red men rose to their feet in angry protest. Chief Flying Eagle then stalked across to the open space where the speaker stood, surrounded by important dignitaries and others. Absolutely ignoring the entire group, Chief Flying Eagle turned to the microphone and thundered in excellent English:—

"The white speaker is wrong! To all who now hear I say: Some white men have seen Sasquatch. Many Indians have seen them and spoken to them. 'Sasquatch' still live around here. I have spoken!"

The chief then strode back to his place and signed to the other Indians to sit down, leaving behind him the Government spokesman whose face was exceedingly red!

I was one of the party gathered about the microphone and immediately said a few words over the loud-speakers to appease the angry Indians. I corroborated Chief Flying Eagle's statement that white men have seen 'Sasquatch,' adding that, although in sadly reduced numbers, 'Sasquatch' are still believed to inhabit the vast mountain solitudes of unexplored British Columbia.

To read Burns is to get the last taste of the old frontier. But, sadly, even by the 1920s facts were too tenuous. Lines between animal and man eventually blurred even amongst the Indians. By the 20th century they were willing to regard most any report of a hairy bipedal creature as a "Sasquatch." But their history tells us something different. In artwork and accounts it indeed tells us of 2 tribes, one portrayed ape-like and the other a very wild and different type of human.

This reality is long buried in America, obscured by the frivolous. In Russia, however, the same phenomenon had also long been around. But because the areas still remain wilderness territory— from the Caucasus to Siberia— the stories remain intact, aboriginal and preserved. Russian studies increasingly draw not a rosy picture of a furry friend but an increasingly grotesque profile of a degenerate and hideous enigma parallel to nothing our fancies could dream up. Free of the American carnival, we must now begin to unmask the real Sasquatch and Abominable Snowman.

CHAPTER 6

# Anthropus X

While all this nonsense was mounting ever upwards, and in the process warping the North American image of the Sasquatch, an actual scientific study was underway in Russia of the sightings of a "hairy wild man" there. The difference between the American and Russian operations was very profound. Most of the Russian expeditions were anchored by scientists and concentrated in genuine mountainous wilderness, centered mostly in the Pamir Mountains north of Hindu-Kush in what was then the Soviet province of Tadzikskaya. But what was most profound of all was the fact that the Russians did indeed find and eventually cast large genuine human footprints. The Pamirs' geographic isolation rules out hoaxers, as the footprints found there have truly been found in the wilderness.

Anthropologist Myra Shackley describes them:

> The word 'pamir' means a mountain valley of glacial origin and the valleys themselves, generally at altitudes of 12,000—14,000 ft, often contain streams which may feed a series of lakes, and include patches of alpine meadow (in the short summer when the area is not completely snow-covered) which produce excellent pasturage for animals. The valleys suffer, however, from an almost total lack of timber and ground suitable for cultivation. The Pamirs consist of a series

of eight separate ranges, each crowned by peaks over 20,000 ft, seamed with ice fields and terminating in glacial moraines. They are often described as savage, inhospitable and desolate, and are certainly very remote; as late as 1926 a scientific expedition found a village there which had had no contact with the outside world for centuries and had developed quite separately.

World attention was brought to the hitherto unpublicized Russian study in 1958; in that year the encounter of a Russian mineralogist, A.G. Pronin, landed in the Russian newspaper *Komsomol Pravda*. In the article he described being near the Fechenko Glacier on August 12th when he saw at a distance a creature "reminiscent of a man's figure, but with a strongly hunched back. Against the white backdrop it could be seen clearly that he was standing with his legs wide apart, and that his arms were longer than in an ordinary man. I stood there, not moving. And so five minutes elapse. The figure then vanished, hidden behind a rock." Due to the worldwide taste for things Yeti, the British *Guardian* picked up the article on February 28, 1958, linking Pronin's sighting, naturally, to the "Abominable Snowman" of the "nearby" Himalayas. The results were dynamic enough to warrant the Russians revealing that they had already been studying this same possibility.

The details came out in an article in *Komsomolskaya Pravda* on July 7th, 1958, written by Dr. Boris Porshnev. The article introduced him as none other than the head of the "Commission for Studying the Question of the Abominable Snowman," with the surprising clarification to follow that he was answerable to the Presidium of the Academy of Sciences of the U.S.S.R., meaning this was also an officially endorsed study.

Porshnev's article showed how long he had been studying the question. Pronin's sighting constituted only the last in an impressive database. "Report" is a term often applied to his material, but that is a misnomer. His "report" is really a synthesis of an accumulation of the reports and data of several investigations. As a doctor of History he was the first to think to look into this inexhaustible vault, discovering forgotten historical archives which had been lost during the Russian revolution. The "wild man" had actually been studied with great interest prior to World War I and the Communist Revolution, during which, naturally, the scientists were either killed or seconded to other duties, and the interest subsequently never rekindled. His contribution was the conclusion, which was aptly rendered in his paradoxically dry title— *Soviet Ethnology*— which immediately introduced his theory the "Snowman" is indeed *a form of man* or, as he called it, a "relic hominid;" to be explicit: a living Neanderthal.

There are several paths that led Dr. Porshnev to his controversial conclusion. Especially influential on him was the work of two eminent Mongolian naturalists, professors Jamsarano and J.R. Rinchen, on the subject of a "wild man" living in the area of Dzungaria and Mongolia called the Almas.

Jamsarano (1880-1940) was a famous Mongolian who fell out of grace with Mongolia and then later from Russia. He was first exiled to Outer Mongolia (literally) by the Tsarists, and then sometime after the Revolution was exiled to Leningrad (because of nationalist views). A *damnatio memoriae* had been accorded him by the Russians, and by Porshnev's time he officially had never existed. During his lifetime, however, he was one of Mongolia's great naturalists. He spent a great deal of time between 1899 and 1928 pursuing the Almas question alone. An artist by the name of Soeltai accompanied him on his travels and each time they spoke with an eyewitness Soeltai drew a color portrait of what the Almas had been said to look like. All this was in Jamsarano's archives and apparently lost, as nothing regarding the Almas appears to have been included in the archives in Leningrad where 148 items of paperwork cache are listed under Jamsarano or in Ulan Ude where 31 items were listed under his name in that archives.

Without this data, Porshnev was dependent on J.R. Rinchen, one of Jamsarano's spiritual students and later one of the greatest intellectuals in Mongolia. Porshnev professed complete reliance on Rinchen's description of the Almas, and even took it to be definitive of Neanderthals.

"The inhabitants of the Gobi Desert gave eye-witness accounts of the Almas," wrote Porshnev. "They are very similar to humans, they say, but their bodies are covered with reddish-black hair which is not thick. The skin is visible through the hair, a nonexistent factor in the case of the true wild animals of the desert. The Almas height of body is Mongol. The posture is upright, with a forward bending slant. They walk with knees half bent. The jaws are powerful, forehead low and receding. The brow ridges are prominent in comparison with the Mongolian brow. The females of the species possess long mammary glands. These creatures cannot light a fire."

Furthermore, "Such characteristics as walking upright, and developed mammary glands in the females, sharply denote the variance of Almas from all existing apes, and definitely make them of a nearer approach to man. Anthropology a long time ago established on the evidence of fossil bones that Neanderthal Man also had a bent posture and low-hanging arms, prominent brow ridges, back-sloping forehead, and powerful jaws. Neanderthal Man too walked with slightly bent knees, and certainly no anthropologists influenced Professors Jamsarano and Rinchen, or their informants. Two independent lines of facts deriving from separate research coincided very exactly."

As seemingly radical as this sounds, Porshnev was right in step with the 1950s' rage with cavemen. This had begun in 1950 with the Toriano, Italy, finds. Modern construction workers had broken into a cave estimated to have been sealed for at least 50,000 years. There anthropologists discovered bare footprints impressed in the prehistoric clay. Due to the dates, all were certain these had to be Neanderthal footprints. Thus when scientists working in the Pamirs (for other reasons) sent sketches back to Moscow of the bare footprints they had come across the similarities were immediately recognizable to Porshnev as Neanderthals. Both the Pamir and Toriano types of footprints were wide, with widely spread toes and a big toe that showed it was capable of independent mobility, that is, it was able to extend sideways by about 20 degrees. This is a rare feature in modern humans, but it is something that seemed more common in Neanderthals.

It was really not that novel (at least in academia) for Porshnev to propose that living Neanderthals were still around. Many anthropologists had toyed with the idea of lateral evolution (British anthropologists even did in the case of Yeti and *Gigantopithecus*). Porshnev, in his way, was saying the same thing; that basically Neanderthals were not an ancestral link with man but a lateral relative, descended from some common simian ancestor. In this view, Neanderthals represented the most advanced *homo* of this laterally and independently evolved mankind.

It hadn't yet come to a head, but a prominent anthropologist was preparing a major thesis that modern *Homo sapiens* actually represented *five* distinct surviving evolutionary lines. When Carleton Coon would finally publish his views in 1962 in his *Origin of Races* there was, however, enormous fallout because Coon had proposed that the various races of mankind had evolved to a sapient stage at different times. Furthermore, he posited that the Caucasoid (White Man) did so 250,000 years ago and the Congoid (black African) did so only 50,000 years ago. The reason for outcry at such claims

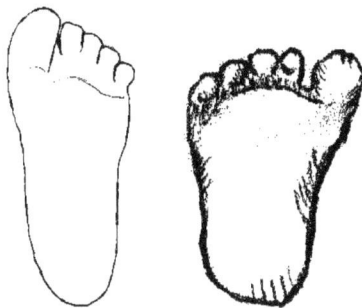

*Two Neanderthal prints, Tana Della Basura (Toriano), Italy, 1950. The latter is much wider, but it shows how the big toe is more offset.*

was obvious. The reaction had a profound effect in evolutionary thinking, burying the very idea of lateral lines of human evolution and separate stages of development.

Naturally, the side-effect wasn't good for Porshnev's views either, which already being associated with the popularized "Yeti" were more or less looked on as a carnival sideshow to begin with. Anybody writing on such topics was and is always easy for an establishment to dismiss. Coon was different. Coon was a Harvard anthropologist and this spoke of established, accepted nomenclature. Everybody reviewed his book, and then screamed for his hair. But for those who had sought the "cryptids" or "hominid relics" there was easy dismissal which never needed to be expressed by a formal established institution. "Oh, they believe in stuff like sea serpents." Porshnev's work also first filtered into English through Odette Tchernine, an English author of Crimean extraction, in her book entitled without any academic air: *Snowman & Company* (1960), something that suggested a compilation of folklore.

Although the establishment stayed away from Porshnev's thesis, it was an enormous influence on crypto-zoology and Bigfootiana. It elevated it to the most profound level possible— a critical pursuit for the very origin of mankind. We must remember that by the time Porshnev's views came to the West, Bigfooters were already sure that Bigfoot was the Eurasian Yeti and thus probably the same thing that Porshnev was discussing.

The problem with Porshnev's thesis, however, was very real in and of itself let alone for how it could be applied to the carnival in America. Serious criticism of his work might have uncovered both its weaknesses and some of its genuine and disturbing implications. But that was not the ken of the woodsmen's craft.

The most obvious weakness was his belief that Neanderthals were

stooping hairy ape men. Only the first skeleton found in the Neander Valley of Germany had a stooping posture. This specimen also
suffered from severe osteoarthritis. This single specimen inspired
the very popular if not universal view of Neanderthals in the 1950s,
and thus such a view was not unique to Porshnev. But it clearly
blinded him to any other interpretation of the Almas data which
was at odds with his thesis.

This is seen poignantly in his misuse of the data accumulated by
a Russian zoologist named Vasili A. Khakhlov. Comparing Khakhlov's data with Rinchen's, upon whose description Porshnev had
so relied, impeaches the whole idea that the Almas and Neanderthals are one and the same.[31]

In 1907 Khakhlov was a young zoologist whose curiosity was
whetted when he had heard of "wild men" in the region of Dzungaria near Mongolia. When in 1910 he had heard about more of them
living in the Pamirs called the "Almista," he made a report to superiors. One laughed it off, but the other, a professor P.P. Sushkin,
who was later to become a well-known Russian scientist, encour-
aged him to study it further. Khakhlov abandoned his university
work and spent years studying in the Zaysan and Tarbagatay
mountains near Kazakhstan. During this time, he found two unrelated Kirghiz tribesmen who had seen specimens of the wild men.
Both were placed in interrogation separately (to keep their stories
separate since neither had ever met). Author Odette Tchernine,
who translated much of Porshnev's material, summarizes the surprising information that Khakhlov achieved from one of the Kirghiz who had temporarily captured one.

The species was a male, of less than medium human height, and covered
with hair, 'just like a young camel,' said the local herdsman. [The Central
Asian dromedary of Mongolia and China with two humps, has thick reddish winter fur, rather like a teddy bear.] But what was really striking about the captive was his long arms, reaching to below the knees. He was stooped, with
shoulders bent forward, and had a narrow, hollow chest. His brow was sloping,

---

[31] Myra Shackley, studying the question much later (in the 1970s and 1980s), was uncertain
whether Porshnev had actually ever turned up the Jamsarano archives and consulted them.
In his own published works (1963 and 1968) Porshnev's notes indicated he was unable to
find the archives or that they had been destroyed. It also doesn't appear that Rinchen did
more than exchange some letters with Porshnev and perhaps later met him briefly. According to Shackley, the Jamsarano archives do survive but remain specifically unlocated. If
Rinchen knew about it, he may have been anti-Russian and therefore did not give Porshnev
any detailed information.

and the brows jutted out sharply above the eyes. His lower jaw was massive, and he was chinless, the nose small, with big nostrils. He had large ears with no lobes, and somewhat pointed towards the rear, 'just like a fox's.' The skin on his forehead, forearms, and knees was horny and calloused, his legs far apart and bent at the knees. The soles of his feet resembled human feet, but were one and half times or twice as broad, with widely set toes. The big toe was shorter than in a man's, and set apart from the other toes. The hands, with their long fingers, had some resemblance to a man's. A curious feature was that he had a protuberance at the back of the neck, rather like some dogs have.

Porshnev must have leapt at that last sentence. He interpreted that "protuberance" to suggest a protuberance that exists on some "Classic" Neanderthal skulls, a feature some anthropologists consider a counter-balancing weight for the long, heavy skull. The description of the foot was then taken to be the Neanderthal foot (Tana Della Basura). "The big toe was shorter than in a man's, and set apart from the other toes."

Khakhlov's description of the "Almista" is not at variance with the Mongolian evidence and even adds the detail about the shape of the ear. Therefore it is hard to see how the Mongolian Almas or any such creature as this could ever be considered human or even subhuman; and certainly there is nothing to suggest a Neanderthal, even as conceived in the 1950s' days of illustrative imagineering. The then-popular and -scientific impression of Neanderthals was that of hairy, heavy set brutish ape men carrying a club and dragging off women by the hair. The Almista description above, with thin-chest and paw-like feet, pointed ears and excessively long arms, is that of a strange humanoid monkey. In fact, there are a number of monkeys noted for having such ears. Instead of suggesting Neanderthals, the Kirghiz seem to be describing a class of primates neither men, apes or monkeys, but a mixture in many ways of all three.

Certainly such a statement can be disputed, but it is interesting to note that there is ancient evidence such an agile monkey-like bipedal primate has long existed. It seems too coincidental that Khakhlov's Kirghiz tribesman should describe something incredibly reminiscent of Greek descriptions and artistic renderings of pans, fauns and satyrs, the pointed-eared bestial if not playful wild men of the mountains who followed herds and kidnapped women. In Illyricum in 86 BC a "sleeping satyr" was even captured by the Roman troops of Lucius Cornelius Sulla and brought before him. The "satyr" was described as brutish, and its language "something between the neighing of a horse and the bleating of a goat." J.R. Rinchen also uncovered this trait in the Almas. Peoples in Dzungaria, Mongolia and Siberia regarded the "wild man" as being mute but capable of laughter. Not laughter as a human, but more like a hyena and other animals that make a whinny, warbly sound. . .Coincidently again, does this not remind us of Roe's description of the Sasquatch's language?

Folklore tells us about the wicked Pan. He inhabited the high hills and was interested in flocks and herds. If we accept this was inspired by such things as Almas in ancient times, we can say that the fact behind the myth translates down to a mountain dwelling hairy primate that was often molesting or just following the herds.

On the island of Delos off the coast of Turkey there also exists a statue group showing Pan, Aphrodite and Eros. Here the short, wicked Pan looks up with a sinister, lewd joy at a naked Aphrodite. There is much symbolism and embellishment in Pan: he has a faun's hind legs, a "man's privy parts," goat horns, pointed ears and, here, the artist rendered with admirable skill the face of a strange ape, with a chin and a short camel-like muzzle. Where did the artist, who carved this around 100 BC on an Aegean island, get so clear a representation of a strange pointed-eared ape? Is it from ancient "Almas" when they still inhabited parts of Europe?

Images of Pan vary, but he is often shown with a heavy brow and thick, protruding lips— features inconsistent with those many goatlike attributes that are in symbolic juxtaposition on the old idol. He was to the ancients a "goat man." For modern man, possessing a knowledge of apes, we might call him something like a "monkey man." Similes aside, all descriptions approximate to the same thing, and that is rather impressive.

Legend had it that in the 20th century such a creature still dwelt in the Caucasus Mountains, not far from the Black Sea and therewith juxtaposition to contact with ancient Greece and the Aegean. In 1959 one Russian scientist pursued the evidence far enough to have an alarming encounter. Yuri Ivanovich Merezhinskiy was holder of the Chair of Ethnology and Anthropology in the University of Kiev. Along with a couple of others, Merezhinskiy was able to penetrate high up into the remote villages. With him was medical doctor and mountain climber Jeanne Josefna Kofman, who would become one of the legendary Soviet Snowman Commission members. Because of her account it is possible to confirm that the

*Pan, in the statue group from Delos. Because the Chimpanzee so suggested a short, hairy little man, zoologists actually gave it the scientific name Pan troglo-dytus. Was Pan, however, based on an Almas?*

with that of the ancient satyr and pan of nearby Europe and with that of the Almista of the Pamirs and Almas of Mongolia.

According to Kofman, they continued to follow lead after lead referring to the *kaptar* until they came upon an old hunter, Hajji Magoma. He agreed to take them up to a hideout to see one, insisting first that Merezhinskiy promise he would only photograph it and would not shoot it. This was a special *kaptar*, Hajji said, for it was a white kaptar. Magoma said there were many who did not believe in them and he would lead them up there to finally put to shame the pessimists by allowing it to be photographed by a responsible scientist.

Merezhinskiy quickly agreed. . .but also as a precaution secretly took along a concealed pistol. On the night of September 18, 1959, they were at the location near a watering stream in the mountains. The night was lit with a half moon, and from their sequestered location they were able to scan the river, shimmering silver in the moonlight, where the *kaptar* was said to come down to drink.

After a while they heard splashing. Concentrating their vision on the stream, they noticed something bathing in it. Then the creature emerged on all fours and stood up. It was thin and white, with thin, long limbs. It immediately started to make sounds of a whinny and warbling laughter. Hajji whispered quickly "snap him." But instead a blast ripped passed his ears. Merezhinskiy had, on compulsion, fired at the *kaptar*, missing however. The creature ran off. Merezhinskiy was so shaken that when Kofman arrived from a different viewing position she reported he was sweating, shaking and facing Hajji's scowl.

Until Kofman could ingratiate herself, no outsider would be led high up into the Caucasus. In the mid 1960s, she was finally able to set up a base camp, and over many expeditions, often into areas where they were closed off by mudslides, she was able to find solid evidence. In March 1966 she described her findings, including having found the lair of one of the *kaptars*. In it were 2 pumpkins, eight potatoes, a half-chewed corn cob, two-thirds of a sunflower center, blackberries, and the remains of 3 apples. To clinch the association, 4 pellets of horse dung were found. The local people had said the *kaptar* eats it for its saltiness. Searching cornfields revealed where a *kaptar* had tasted cobs to find the sweetest. It left tooth marks proving it had a wider mouth than humans. This was something the local people also confirmed. It was said a *kaptar*'s massive teeth are able to crush even a squirrel's skull. It is not fond, however, of the fur.[32] Collating this with the Pamir Almista data does seem to confirm the Almas/*kaptar* had a wide habitat.

Amazingly, Porshnev accepted all this as suggestive of Neanderthals because to date (1966) only one type of footprint had been sketched, and these were those that resembled the Toriano Neanderthal finds. Yet eyewitness descriptions in both the Pamirs and Caucasus continued to point to something far more animalistic and consistent with ancient accounts of giggling pans.

The Moslems of the Caucasus even referred to such a hairy creature by a word familiar to almost anyone in Western Civilization— Shaitan. . .*Satan.* It is interesting to note here that in the Jewish ritual of the Scapegoat, the goat is released to wander out into the wilderness, delivered unto Satan. The Hebrew word is actually Azazel. Rabbinic scholars preferred to translate this word as a rocky wilderness place, but they were well aware that Azazel had been popularly personified by the ancients as a hairy, faun-like monster of the wilderness just like Pan. The picture conjured by this association is that of Pan taking sheep or goats released into the wilderness, the place of his abode. One cannot help but wonder how such divergent and opposite religions like the polytheistic Greeks and the monotheistic Jews could have a similar concept of something "wicked" like Pan that inhabits the rocky places and mountains, preys upon herds and terrifies people, if there was not some unique creature behind it.

A similar legend was maintained in the far-away Pamirs. In America we are used to the names that the Indians gave such "hairy wild men," like Sasquatch, but we employ them merely as monikers with often-misunderstood meaning since White Man's language heritage is not the same. An example, however, of how a hairy manlike creature has been constantly regarded in Eurasia can be found in the various names given to it by the local peoples since they also share a connection with our own Indio-European language root. One name is the *Dev*, from which our own word devil is derived. It is regarded as an evil spirit, of course, a harbinger of bad luck, but in the Pamirs it is also applied to a hairy manlike creature when encountered.

One encounter with a *Dev* was had by zoologist B.M. Zdorik in 1934, in company of his Tadjik guide when they were in the brush of the Darvaz Range, near the westerly spur of the Pamirs called Peter the First. He wrote a report, containing the following surprising narrative: "Suddenly, a small area opened up in front of us on which the grass was completely flattened, and the ground dug up as if it had been done with a spade. On the path were drops of blood and scraps of what looked like marmot fur. But there, right at my very feet, on a heap of freshly dug earth, an unknown creature lay asleep. It was lying fully stretched out on its stomach, about 1 ½ meters or so in length. I could not see the head and front limbs very well as they were hidden from me by a withered grechika bush. I did manage to see the legs and the bare black feet which were *too long* and too well shaped to be a bears."*

The creature's flanks, wrote Zdorik, rose and fell rhythmically as it peacefully slept. Its hair was a redder color than he had ever seen on a bear. It was also different, looking more like a yak's wool than the downy look of a bear. He admitted that he stood there "frozen with surprise, and at a loss as to what to do." He looked behind him to his guide, who stood with a face as white as a sheet. "Never before had I seen such an expression of terror on a man's face. His

---

* Italics mine.

fear communicated itself to me, and beside ourselves, without glancing backwards at the creature, we both fled away down the path, enmeshing ourselves and stumbling about in the high grass."

Another excellent example of a "Pan" far from Europe comes from a zoologist, K.A. Satoonin. In the late 19th century he encountered, as he described it, a speechless female of a type of hairy manlike beast. Satoonin then probed more deeply into the existence of the Almas, and published a paper in 1899.

Something similar seemed to follow the reindeer herds as far away as Siberia. Archeologist A.P. Okladnikov wrote a report concerning his experiences north of Mongolia along the Lena River. It is contained in the Porshnev papers, and holds the account of wild men called by yet another name. "The Chuchuna," he writes, "are a tribe of half-men, half animal beings, still occasionally met with in the North. The creatures have no neck, and heads that consequently seem to sprout straight up from their torsos. They usually appear at night, unexpectedly, and throw rocks on the sleeping humans from the cliffs. They are given to trapping reindeer. A Yakut hunter named Makarov said he found caves inhabited by the creatures on the River Lena's right bank and as far as Lake Stolb. In these lairs were many antlers, and some hides of the reindeer that have been eaten."

Khakhlov's concept of the Almas

Despite all of these accounts, spread over the last century, the "wild man" reality remained a source of derision. In another case along the Lena River a report from mineralogist P.L. Dravert was filed in 1912 describing wild hairy men who were incapable of human speech. In 1933 he published a long paper detailing the Mulen, as they were also called, only to be written off as a crank.

The few Russians who were investigating about 100 years ago also seemed to regard the Almas as some form of subhuman long before Dr. Porshnev took up their lead. In Khakhlov's case, he even showed his two witnesses pictures of gibbons, gorillas and chimpanzees, and then showed them a picture of a reconstruction of a prehistoric man, as science then fancied he looked. Both pointed to the caveman as the closest thing resembling the Almas. As a result Khakhlov suggested the scientific name *primihomo Asiaticus*— "first Asian man."

From this it is certain that Khakhlov thought he had something far more primitive than Neanderthals or even *Homo erectus*. When he sketched his impression of an Almas, he drew an ape cone head, a trait ostensibly inspired from the mere mention of a "sloping head." No caveman reconstruction ever showed such a trait. Nor would one show a "fox's ears." Khakhlov's sketch looks like a flat faced monkey with a pointed head. This is nothing human and certainly not Neanderthal.

Somewhere between man and animal is where most of the indigenous peoples of Eurasia seemed to place the "wild men." The curious position in which the medieval Mongolians held the Almas highlights, once again, the dichotomy of the subject. In the 1950s, anthropologist Emmanuel Vlcek discovered two anatomical and medical books in the Gandan monastery in Ulan Bator. One, published in the late 18th century, had a large section devoted to the fauna of Mongolia, including amongst its many illustrations of wildlife the "wild man." The text reads: "The wild man lives in the mountains, his origins close to that of a bear, his body resembles that of a man, and he has enormous strength. His meat may be eaten to treat mental diseases and his gall cures jaundice." No more clearer indication can exist than this that the "wild man," despite the name, is not regarded as human, for no Buddhist would eat human flesh for medicinal purposes.

Myra Shackley writes in response:

> It is particularly interesting that this creature, which is so clearly an Almas, must have been sufficiently well known to travelers between Tibet and Mongolia to have been included in what is really a standard work on Mongolian natural history, as it may be applied to Buddhist medicine. The book contains thousands of illustrations of various classes of animals (reptiles, mammals and amphibian), but not one single mythological animal such as are known from similar medieval European books. All the creatures are living and obtainable today. There seems no reason at all to suggest that the Almas did not also exist, and the supporting text and illustration seem to suggest that it was found among the rocky habitats, in the mountains.

Ancient artwork in both Tibet and China also preserve the existence of just such a human-like primate. On Tibetan *mandalas* there appear any number of animals that are attendant on gods or are aids in meditation. The "wild man" appears on here quite separately from bears, langurs and apes, making it quite apparent that the ancients had no trouble in separating them from animals and

*The Chinese and Tibetan* bitchun *or "wild man."*

people. Occasionally the wild man also appears on Tibetan *thang-kas*, placed in the cosmos below humanity but above the animal world.

Although the wild man definitely earned his place above the animals, it is nevertheless true that based on text and not pictures alone the wild man was regarded as an animal. There is a case in the Porshnev material that justifies this. In 1959, a soviet militiaman, Mattuk Abderaim, was on vacation to his uncle's in Tashkurgan, near the Pamirs, when his uncle returned home with a killed Yavo Khal'g, a variant name in the Sinkiang Province for the wild man, also known as Yaboy Adam or Yabalyk-Adam (Adam meaning man). Young Mattuk Abderaim was familiar with what apes and monkeys looked like based on books he had seen while in the military. However, he said this thing was more manlike except its thumb was jointed like its fingers (in other words, not opposing), *a trait only of monkeys.*

The most complete description of an Almas comes from General Mikhail Stepanovich Topil'skiy, and this shows us plainly why it held such a provocative position between man and animal. During the Russian civil war the general was in charge of men who were pursuing a renegade band of Whites (the royalists) in the region of the High Vanchsk. They had already been forewarned about "beast-men" living high up in the area. There was little worry about encountering one around the villages; the beast men kept to the mountains and were only known now by their strange wolf-like cries coming from the high altitudes. It came about, however, that the general was forced to investigate the mountainous area in search of the Whites. In doing so, there followed a number of small skirmishes with the Whites, and in one instance his men fired into a cave believing that Whites were in there. What happened came to the general's ear shortly after from a wounded soldier.

The soldier reported that a wild hairy man making inarticulate sounds came running out and was killed in the machine gun fire. The general was skeptical. But considering the legends that the local people had told him, he decided to go along with the medical officer and investigate. The general reported: "At first, I thought it was the corpse of an ape. It was covered with fur. But I knew there were no apes in the Pamirs, and, moreover, the body looked far more human than ape-like, indeed, fully human."

The medical officer established it was definitely not a man as we know them. They rolled it over and measured it. It was close to 6 feet long. The hair, which was all over its body, was mostly grayish brown, with gray hair mostly covering the belly. The creature seemed to be old. The distribution of the hair was unusual but followed an alarming pattern: on the chest it was thin, but on the belly the hair was shorter and thicker. The hair all over was very dense, yet without an undercoat. The skin was plainly visible underneath. The thighs and legs were the hairiest of all. There was no hair on the knees, which were calloused, and very little on the buttocks, causing the doctor to conjecture that it sat like a human. The shins were less hairy and the entire top of the foot was hairless. Shoulders, arms and forearms were covered with hair, until around the hands where the hair was very thin. The palms of the hands were also hairless. The neck was very hairy, unlike a man, but there was no hair on the face, like an ape. There were only a few hairs on the upper lip. There were no eyebrows. This area was a thick, jutting brow like in apes. The head sloped toward the back, with no hair on the low forehead above the brows but becoming thick and matted behind the head like apes.

The 'thing' was lying dead with its dark, beady bear-like eyes staring. Its teeth were plainly visible in its death grimace. They were thick and even, like a man's. It had high and pronounced cheekbones, a flattened small nose with large nostrils and a deeply set bridge between the eyes. The ears were hairless and slightly pointed at the top (like Khakhlov reported) and, in this case, the thing had longer earlobes (though in other accounts they have none). The lower jaw was jutting and huge, unlike an ape which has a muzzle.[33]

The chest was massive, strong and broad, and the muscles of the arms were impressive and powerful looking. The torso was similar to a man's, narrowing at the waist. The hands and feet were broader than a human being's. They were thickly calloused. The skin was a dark brown.

With no other choice at the time the soldiers buried the freak under a pile of stones. What was it? The general does not mention that the arms were longer, only that the fingers and toes appeared in normal proportion but that the feet and hands were shorter and wider. The identity with the Almas is noteworthy except for this last point (and the broad chest).

Perhaps it is an oversight or perhaps the "thing" became more human looking in the retelling. It is hard to imagine how Topil'skiy could have overlooked the unique paw-like foot of the Almas as described by the Kirghiz to Khakhlov, or the 5 fingers young Mattuck reported.

Time and sensationalism may indeed have something to do with eradicating these unique Almas features. In 1933 a geologist named D.I. Shcherbakov would have an encounter in the same general area which would fill in one notable blank left in Topil'skiy's report. It was near the Yazgulem village at the Pamirs while crossing the same Vanchsk uplands to the Yazgulem elevation. He found naked footprints and was surprised that they resembled neither a man's tracks nor a bear's. The big toe was the major problem; it bent out along the side. Adding this to the pointed ears on Topil'skiy's wild man and we seem to have the Almas as described to Khakhlov.

But, alas, there is no question that great ambiguity surrounds the Almas. Okladnikov and Dravert's accounts highlight the confusion. Dravert made it clear he is describing people called Mulen and Chuchuna. Okladnikov is describing a giggling Almas. Dravert reported that in other parts of Siberia the Chuchuna are said to wear clothes and use bow and arrows, not a trait of a "Pan" or Almas.

The above accounts underscore the problem that an animal is called a "wild man." But the true controversy is that on occasion

---

[33] Note the similarities to Roe's description of his Ousquatch and the Indian burlium masks representing the snarling "wild man of the woods."

men have been confused for these same hairy primates. This could not happen if the humans did not look remarkably similar to the Almas and more apelike than we do. Therein is the great controversy and therein is where Porshnev got most of his inspiration to believe that *all* Almas were really Neanderthals.

Examples of animalistic humans include the interesting legend in Tibet and China that says that the Monkey King and the female rock dweller (Almas) were ancestors of some of the people. This has particular relevance to Porshnev's views in light of English explorer Gordon Creighton's observations. In 1935, Creighton was in Peking participating in a function that brought together many of the Tibetan Lamas and Sir Charles Bell, the Tibetan expert. He noted how many of the Lamas had notably high-domed heads, as the Mongolians insist the Almas also possess. The Tibetan entourage of one lama (the Tashi Lama) in the formal picture that was subsequently taken were easily recognizable by their dome-like heads. One lama was even known as the "son of the female Almas." There is no way that the lama's mother was not human, but what or, more accurately, who were *these* Almas?

Ambiguity unfortunately is the fertile field of hominid research. No matter how many names there are for wild men amongst the various nations, "wild man" is all they amount to. It is not surprising therefore that so much gets intermixed and intertwined. Creighton compiled a list of possible "remnant hominids." There were at the date of his writing at least 127 different names. Except for a few variants like the American carnival "Bigfoot" all mean "wild man." Even Sasquatch comes from the wild men of the Saskahaua district. Such an elastic term could easily be applied to both man and a bipedal primate completely unrelated to mankind.

Since ambiguity implies similarity in two different subjects, we thus also come to the great strengths of Porshnev's material. He collated so much data there should not be any doubt on the existence of the animal Almas. But he also gathered enough data that there should be no doubt that *humans* are also involved.

'A wide foot with a shorter big toe offset.' That is the conundrum. Its exact meaning comes down to interpretation, doesn't it? But solid footprints don't. And now we come to the tangible data upon which Porshnev's theory hinged. After his death in the 1970s, the footprint sketches he had long relied upon materialized in plaster. Two members of the "Commission" cast very unusual footprints: Dr. Jeanne Kofman in the Caucasus and Dr. Igor Bortsev in none other than the Pamirs where lived Khakhlov's "First Asian man."

The footprints almost certainly appear to be those of living Neanderthals. In the print Kofman cast in 1981, the big toe is truly at a 20 degree angle to the others and has greater independent mobility. For the print Bortsev cast two years prior in the Pamirs the big toe is appressed to the others but looks capable of bending outward independently as in the Caucasus print. Neither are as paw-like as the Tana Della Basura print, but the big toe is offset in the Caucasus print in a similar way.

I call this last print the "Gool Foot" because the peoples of the Caucasus also mentioned "wild men" that they called Bianbangooli (forest ghoul). A variant of this same name is used in the Pamirs (Gulbiyavan). I make the distinction here for the footprints because

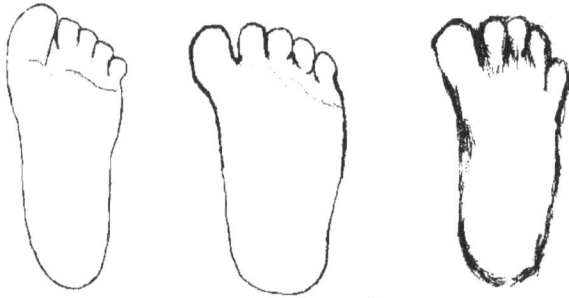

*Neanderthal, left, Toriano, Italy, 1950, and the "Gool" of the Caucasus Mountains, 1981, found by Dr. Jeanne Kofman. Print found by Dr. Igor Bortsev in the Gissar Range of the Pamirs, Tadzhikstan on August 21, 1979, reveals the Eurasian connection.*

it seems locals do as well. Specifically, the "Gool" appears to be an entirely separate entity to the Almista and *kaptar*, and neither of these latter names are ever applied to the Bianbangooli. In fact, Almista is never used in the Caucasus for anything; nor is *kaptar* used by the peoples of the Pamirs. But both preserve the same concept of the forest ghoul. And it may be the existence of these "forest ghouls" that confused Porshnev to think all Almas were some kind of subhuman. Unfortunately, he always (and thus the Soviet Snowman Commission) made the great mistake of calling *everything* an Almas. In a later interview (in the 1980s), Kofman even said she called every report of a hairy biped Almas "because Porshnev did."

Another footprint discovery in 1972, however, provides us with equally tangible data (and far more precise to the point) showing that two entities are and have been involved. Just such a foot as the Kirghiz described to Khakhlov for the Almas was unintentionally found. This was during the Cronin-McNeely biological expedition in the Himalayas. While in the Arun Valley of Nepal, Edward Cronin and Jeffrey McNeely's camp was visited at night by a curious bipedal creature with remarkably ape-like but paw-like feet. This was discovered upon opening their tents in the morning and seeing that it had meandered about and investigated their camp. Following the tracks confirmed it was bipedal. They came down a ridge, without any sign of the assistance of hands or forelimbs. Being there for scientific purposes, the team had plaster and cast one of the prints. Photographs of the others show a very mobile big toe. But the cast, taken of the foot when the toe was held appressed, shows a paw-like foot with an offset big toe shorter than the others, just as the Almas foot was described to Khakhlov in 1910.

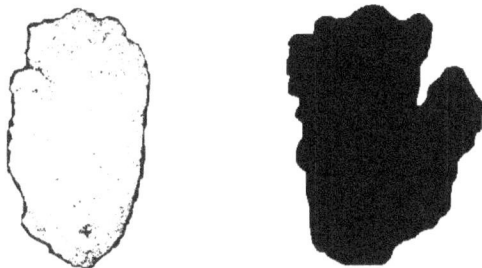

The McNeely Print from 1972. Left, an outline of the cast (with toe held close); Right, an outline of one of the prints in snow.

The footprint is clearly not a Yeti, though like the strange footprint Tombazi records it treads the Yeti territory. It is not important right now to discuss how one creature might migrate and tread the territory of another in doing so. It is enough that a footprint has been independently established to suggest the Almas is real and not mythological, and, more importantly, is *not* a human being. The McNeely Print no doubt belongs to that strange hand young Mattuk Abderaim saw on the dead Yavo Khal'g, that monkey hand of five fingers and no thumb.

However, the controversy of the Russian material cannot be redirected by this discovery onto the path of only a genus of manlike "monkey anthropoids." There is left to be answered those large Neanderthal type "Gool" footprints, for they are the ones that have openly placed humanity within the discussion. They were cast in the remote wilderness indicating a people we thought not possible, then extinct and now in modern times rationalized to merely a unique variation of ourselves, *Homo sapiens.*

Thus we cannot avoid coming back to the controversy, which rather heightens here, for there are no descriptions of normal naked, wild people, coming from the Caucasus or Pamirs. There are only descriptions of a hairy beast man. In the former by the Black Sea, precisely where Jeanne Kofman concentrated her studies, such wild men were historically considered to have been an actual tribe of people, much like the Salish Indians regard 'Sasquatch men.' Here, too, there are those historic stories of capturing one, and within this category there is a disquieting case that proves humans *are* involved.

The famous case of Zana, as she was later named, is well-known but very infrequently has it been told accurately. Dr. Porshnev actually personally journeyed to the wilderness area where she had been captured and domesticated to search for her remains. He was sure that if found they would prove his supposition that Neanderthals were still living. His work alone constitutes the best and most accurate recounting of this bizarre story, and it is fortunate that it exists in his material and that people like Odette Tchernine were able to translate it.

Dr. Porshnev first heard of the story in 1962 from his Snowman Commission associate, Professor A.A. Mashkovtsev. In the area of the Caucasus they called such wild people *Abnauayu,* which tribe it was said in times past the Abkhazians drove out into the mountains when they settled there. Zana was considered to be a remaining female from this ancient tribe. She was captured sometime in the 1840s or 50s and then died in the 1890s. As long after as 1963, however, there were still at least ten people living who had been at her funeral, and she was remembered by 100 or so in the village of Tkhina where she had lived and was then buried.

As to how she was captured there was some doubt. Some said she was found in the woods of Zaadan. Others said she was found on the sea coast of Adzharia. Most believe the latter, since there might be some connection with that and her name of Zana which in the Georgian language means dark-skinned or negroid, which fit her description. In any case, Zana was captured by hunters in a way that is merely described as an "age old technique." This may have been on behalf of the local ruling prince D.M. Achba.

When they first captured her she fought back violently. They had to subdue her by binding her with cudgels, gagging her and then shackling her legs to a log. After being presented to the prince she was given to one of his vassals, Kh. Chelokua, and then later taken away by a nobleman to whom she was later presented. Then Edgi Genaba took her to his farm near the village of Tkina. How many years had passed is not mentioned. Perhaps it was not many, for when Genaba bought her she still had to be transported chained up due to her furious and violent nature.

At her new home on his farm she still had to be locked up in an enclosed kennel or blockhouse, and food could only be thrown at her. Eventually, as she grew tamer, she was let into a corral, still tethered. Then after some three years of captivity she was tame enough to be allowed out to wander around.

Now that she was free to roam, she acted more or less like a domesticated but still slightly wild mental defective. I assume she was tamed by numerous beatings, for she developed a most unusual dread of her master and would quickly cower when he would shout at her. She slept outside since she could not stand warm rooms. She would dig a hole and curl up in that.

Those who still recalled her told Dr. Porshnev that there was nothing human in her expression whatsoever. She had a beast's features. More than one said her face was "terrifying"— something we can well imagine from their description. She had dark brown skin, or light black, red tinted eyes, high cheekbones, flat nose, strong teeth that could crush anything, and she was covered in reddish-black hair, with it tousled and full on the head like an old furry hat and then long down her back. She had full and large breasts, thick muscles on her arms and legs and she had long, thick fingers. All these traits came together to form an expression that was devoid of anything but bestial menace.

Zana ate nothing but raw meat and nuts and fruits, and was trained to do simple slave-like chores around the farm, like grind grain. She liked to sit and play with rocks and even chip them together. She also liked to throw them and when teased by children she threw them at them and chased them off. For years and years she lived like this, seemingly content. She would never wander far from her master and easy food. When taken into town she inspired dread and fear in most of the people. Dogs couldn't stand to be around her, and she in turn hated dogs.

So what was she? Was she a real Almas or Almista, or were the *Abnauayu* of the Caucasus something else?

This must be asked because Zana did in fact have children. Apparently she was not too frightening for some of the lesser and more desperate blokes of the village, and she conceived 'half-breed' children. She needed no assistance in birth, and afterward she also took the children to the stream to wash them off, drowning them accidentally in the process. Finally, the townsfolk intervened, and when she conceived and gave birth again they seized the babies and raised them in town. Altogether she had four surviving children, two girls and two boys. Their names were: Dzhanda (eldest boy), Kodzhanar (eldest girl), and then the two younger, the daughter Gamasa, and the last boy was Khvit. The last two might have been sired by Genaba himself since his wife raised both in their own house. Khvit died in 1954, and Dr. Porshnev was able to get a good description of him.

Zana's children were entirely human, though darker and with negroid features to their faces. Khvit had been unusually strong. He was capable of social interchange and holding down a job, but he would be more difficult, wild and hard to get along with. Khvit did not have kinky hair: it was straight and soft, but he had thick lips and dark skin. The other children had some similar physical and mental eccentricities, but were regarded as fully human.

Of Khvit's children, Dr. Porshnev met two in 1964 at the village of Tkvarcheli. He recalls: "From the moment I saw Zana's grandchildren, I was impressed by their dark skin and slightly negroid looks. Shalikula, the grandson, has unusually powerful jaw muscles and he can pick up a chair, with a man sitting on it, with his teeth, and he can dance. One of his gifts is imitating sounds of wild and domestic animals."

Pray, just what was Zana? If she could have human, though to some extent eccentric, children she could not have been an ape of any kind. She was no *primihomo Asiaticus* or giggling Almas. She had to be a human being: a wild, ignorant human capable only of grunts and simple chores. She could never learn speech, although her children had no problems.

We come back to the nagging question of more than one "tribe" of "wild men." Of the wild Sasquatch woman he encountered, Charley Victor reported: "The hairy creature, for that was what it was, walked toward me without the slightest fear. The wild person was a woman. Her face was almost negro black and her long straight hair fell to her waist. In height she would be about six feet, but her chest and shoulders were well above the average in breadth." Victor described her expression as fierce and savage, her eyes dark and fiery; very similar to Zana. The difference was that the Sasquatch woman could speak something like Douglas. The Salish human shaman mask of the "wild man" also shows strong cheekbones and very thick lips, just like Zana's children. Underscoring the contention that these represent a type of people and are not isolated or mythological events is yet another discovery nearby to the Caucasus Mountains in Dagestan in 1941. A Russian medical doctor, a Colonel V.S. Karapetyan, was called upon to examine a strange wild man that a military patrol had captured. He wrote a full report which was later disseminated by the Russian Information Service (contained in the Porshnev papers).

From October to December of 1941 our infantry battalion was stationed some 30 kilometers from the town of Buinaksk. One day the representatives of the local authorities asked me to examine a man caught in the surrounding mountains and brought to the district center. My medical advice was needed to establish whether or not this curious creature was a disguised spy.

I entered a shed with two members of the local authorities. When I asked why I had to examine the man in a cold shed and not in the warm room, I was told that the prisoner could not be kept in a warm room. He had sweated in the house so profusely that they had had to keep him in the shed.

I can still see the creature as it stood before me, a male, naked and barefooted. And it was doubtlessly a man, because its entire shape was human. The chest, back, and shoulders, however, were covered with shaggy hair of a dark brown color. This fur of his chest was much like that of a bear, and 2 to 3 centimeters long. The fur was thinner and softer below the chest. His wrist was crude and sparsely covered with hair. The palms of his hands and soles of his feet were free of hair. But the hair on his head reached to his shoulders partly covering his forehead. The hair on his head, moreover, felt very rough to the hand. He had no beard or mustache, though his face was completely covered with a light growth of hair. The hair around his mouth was also short and sparse.

The man stood absolutely straight with his arms hanging, and his height was above the average— about 6 feet. He stood before me like a giant, his mighty chest thrust forward. His fingers were thick, strong, and exceptionally large. On the whole, he was considerably bigger than any of the local inhabitants.

His eyes told me nothing. They were dull and empty— the eyes of an animal. And he seemed to me like an animal and nothing more.

As I learned, he had accepted no food or drink since he was caught. He had asked for nothing and said nothing. When kept in a warm room he sweated profusely. While I was there, some water and then some food was brought up to his mouth; and someone offered him a hand, but there was no reaction. I gave the verbal conclusion that this was no disguised person, but a wild man of some kind. Then I returned to my unit and never heard of him again.

W.H. Tilman, too, recorded a similar description for the "Snowman" in the far-away Himalayas. Recall that Kaulback's porters argued over who made the strange footprints they had seen. A couple insisted the 'mountain men' which they "described as like a man, white-skinned, with long hair on head, arms and shoulders." Charley Victor noted different colors to Sasquatch people. "The old hunter felt sure that the woman looked somewhat like the wild man he had seen at Yale many years before," wrote Burns, "although the woman was the darker of the two. He did not think that the boy belonged to the Sasquatch people, 'because he was white and she called him her friend.'"

The Caucasus and the Himalayas are far away, but no less distant from one another than Siberia and the rocky heights of Eurasia. And here, too, such wild men have been reported which, as we know, Dravert called Mulen and Chuchuna. And from Siberia to Alaska and down to British Columbia the distance is no less or *more* daunting. And here, too, the Indians described, however rarely, similar wild men.

As far apart as these places are, they are no less distant to each other than to ancient Greece and Rome, and the troglodytic caves of the Saskahaua no further in time than the observations of Lucretius of what appear to be living Neanderthals. Certainly something uncannily similar inspired his view of "early man" in his great poem *De Rerum Natura*— "On the Nature of Things":

> But mortal man
> Was then far hardier in the old campaign,
> As well he should be, since a hardier earth
> Had him begotten; builded too was he
> Of bigger and more solid bones within,
> And knit with stalwart sinews through the flesh,
> Nor easily seized by either heat or cold,
> Or alien food or any ail or ilk.

And whilst so many lustrums of the sun
Rolled on across the sky, men led a life
After the roving habit of wild beasts.
. . .As yet they knew not to enkindle fire
Against the cold, nor hairy pelts to use
And clothe their bodies with the spoils of beasts;
But huddled in groves, and mountain-caves, and woods,
And 'mongst the thickets hid their squalid backs,
And trusting wondrous strength of hands and legs,
They'd chase the forest-wanderers, the beasts;
And many they'd conquer, but some few they fled,
A-skulk into their hiding-places. . .
With the flung stones and with the ponderous heft
Of gnarled branch. And by the time of night
O'ertaken, they would throw, like bristly boars,
Their wildman's limbs naked upon the earth,
Rolling themselves in leaves and fronded boughs.

Could a living type of people have inspired the erudite Lucretius' poetic expression? Uncivilized was evocative of "early man" in the Greek mentality, and if Greek naturalists or Lucretius himself witnessed such peoples in the wild they would have naturally thought, as we have too from mere descriptions today, that they were seeing examples of early man, natural, primitive and adapted only to free roaming and hunting.

Our discovery of thick-boned skeletons in caves, of course, does not preclude that the Greeks could not have discovered them as well. The Greeks were great anatomists and, learning from them, so were the Romans. But inspiration for Lucretius' description of this kind of Man seems too close to unexplained peoples still reported in the 19th and 20th century not to reflect some contact with actual living Zana-like wild men rather than just acquaintance with the examination of a skeleton found in a cave and over which the Greek anatomists, like us, also probably brooded.

A connection from Europe to the Saskahaua for this same people is not imaginary. It is bridged in the land of the Almas— the Altai Mountains and Dzungaria, more than halfway to the Bering Straits and the Pacific Northwest. Two discoveries 500 years apart illustrate what has walked this bridge. At the beginning of the 15th century the German traveler Johann Shildberger bore record that there was the belief that beyond the Altai Mountains there was a desert (the Gobi). He then goes on to describe what appear to be the Almas. "In the mountains themselves lived wild people, who have nothing in common with other human beings. A pelt covers the entire body of these creatures. Only the hands and face are free of hair. They run around the hills like animals and eat foliage and grass and whatever else they can find."

Investigating the same Altai Mountains further, in 1979 Myra Shackley came across curious evidence. In fact, it was the only tangible evidence of something having lived in the Altais. "It turns out that the whole area is rich in Mousterian Neanderthal artifacts." The legend of hairy wild men was still intact amongst the local peoples. She noted that no Mongolian goes there because they are scared of the creatures that supposedly inhabit the mountains. "If the Neanderthal were to have survived it would most likely be in exactly those areas where the alma [sic] has been most persistently reported."

Are "Sasquatch men" of British Columbia and the *Abnauayu* of the Caucasus branches of a similar race of men we first discovered in the ancient caves of Lucretius' Europe and have called Neanderthals? Have these troglodytes, for so they seem to be, because of their wide dispersion in the mountain fastness where also lives some kind of unknown primate, been confused for the same thing?

Primitive by our standards would be lack of civilization or city building, as the word means. But if there was a thick, strong people who had a complete intolerance of warm rooms, as it is said many "wild men" have, their world out of necessity would not be one of building cities and developing controlled atmospheres. They would live in the mountains and with time and, with becoming fewer and fewer, become something almost inhuman in nature.

The same scenario of the Altais can no doubt be said for British Columbia. One, human, would leave evidence for its passing (such as in those caves Burns mentioned); the other, animal, has left no tangible trace. . .except in the Indian artwork of *bukwas* masks. Do they tell the truth? Man and animal as "wild men"?

The Russian material is a silent echo of a voice not being shouted in America— "*two tribes*". It does, however, recognize because of Zana that the problem of humans is omnipresent as a completely separate entity. All we have in America are old Indian accounts and our disbelief they can be accurate. The problem has gone largely unnoticed over here because the Neanderthal or "Gool Foot" has never been cast and discussed in America. We are so giant ape consumed that a variant human footprint would probably go unnoticed and ignored, if so that the problem still exists here. It might be more alarming (and certainly more interesting) if such a footprint was cast in America.

Zana could not have been a feral child. Her grandson Shalikula was a living reminder of her strength and power. She was a member of a type of human we thought long extinct, not a delinquent of our own kind abandoned when a baby. [34]

We are thus led to most controversial problem then: that perennial problem of whether "hairy wild men," on occasion, might indeed be hairy wild men, a true Anthropus X. It may not be the maker of the Shipton footprint. It may not be the creature that left the McNeely Print. It may not be the Sasquatch of Roe and Ruby Creek. . .but is there (or were there) "Gools" in America who lived in the same area as an unknown type of bipedal simian?

Charles Stonor's research in the Himalayas underscores the question. Above the 14,000 foot level he and his party pondered strange human prints. The prints were so puzzling that he took a picture. Not only does the print seem to have been made by a bare human foot, the outline of the print (for it had melted out some) matches the prints found by Kofman and Bortsev, and even those found in that sealed cave in Italy. A very specific type of human footprint has been found all over Europe and Asia, where the Yeti and Almas also walk. Is this the footprint of the hairy mountain men Tilman and Kaulback mention?

What a waste California is in the research of this (or these) bizarre and frightening humans and primates. It was a weird interlude of the ludicrous. The Romans called the great games in the amphitheatre the *ludi magni*, or merely "great games," from which we get the term ludicrous— something outlandish and for show. That certainly fits California and Bigfoot. What a waste of research time and effort trying to unsort all the rubbish that the cone-headed "Bigfoot" in California began. Those enlarged comic human footprints were such a shabby gesture. It is amazing that anybody fell for them.

But if the California *ludi* was based on human fake prints, the serious Russian studies have cut to the heart by indicating that humanity *could* be involved, a terrible, frightening and gross degenerate humanity. In a very real way this should resurrect respect for the early Indian stories of Sasquatch men as a tribe of wild Indians who lived deep in the Saskahaua's mountain caves.

---

[34] Snowman Commission member Professor A.A. Mashkovtsev also found a Neanderthal style flint point in the area where Zana had lived before capture.

The publishing of the Russian and Mongolian discoveries in the 1960s should have inspired our own scientific establishment to trek up into the wilderness of British Columbia in order to settle the point. But it did not because of the *ludi* of California. The ironies were obvious and discouraging to a sober approach. For instance, there were few sightings in Russia even 100 year ago. The same can be said for Mongolia and Tibet and Nepal. Yet suddenly in crowded California there is this huge burst of sightings following Jerry Crew's picture with him holding that grossly enlarged human print. This may have resulted in mass interest and a lot of chicanery in America, but overseas the Russians were watching with a shake of a very skeptical head, as Odette Tchernine already in 1970 aptly observed in her book *The Yeti*:

> By contrast the Northern California Bigfoot episodes must be treated very speculatively. Some of the evidence may provide suggestions of the presence of a very primitive human or animal but proof has been dubious down the years, and not only since the recent mystery "outbreak" and the film [Patterson Film]. In spite of the sincerity of many of the witnesses, an atmosphere of contrivance surrounds the story. When a report is repeated at intervals, often treated at entertainment level, and heavily built up to stimulate interest, that is the signal for serious research to stand back on the sidelines and wait. The Californian adventure-getters have spoilt what might have been a rare root of truth by blowing it up into a forest of levity.

The trail in California is a cold one, offering little reliable since the days of the nebulous "Kangaroo Man". But California is on the trail, and America holds a critical piece to the puzzle. The Old World has given us the probability of people, but it has given us only enigma when it comes to the strange anthropoids Almas and Yeti. The New World, however, gives us a clue to their origins. It is best to return to America and follow its clue from a very unexpected place.

# CHAPTER 7

# Odd Anthropoid Out

In the sweltering humid heat of the Venezuelan jungle, Swiss-French geologist François de Loys was leading his weary party of men searching for positive geologic signs of oil deposits. Now in 1920 it had been 3 years since he had led 20 bravos into the "green hell," as the jungles were called. Through disease, snake bites and pitched battles with wild Motilone Indians his party had been culled down by near half.

Now by the Rio Tarra, a tributary of the Rio Catatumbo, something alarming took him and the remainder of his fatigued party by surprise. Two large anthropoids were standing on their hind legs amidst the lush understory before them. Their size was remarkable. They must have been between 5 and 6 feet tall. They were a male and female. Strangely, they looked like common spider monkeys, a species of South American monkey that is 3 feet tall at best, a quadruped that has a long, prehensile tail. Yet it was clear that these anthropoids had no tails and they both advanced on their hind legs like men.

Screaming and howling in agitation at de Loys' party they finally ripped branches off the bushes and brandished them threateningly. De Loys and his men poised themselves for defense. The "monkey" fury continued to a deafening pitch. They aggressively beat the bushes with their sticks, and when their shrieking reached crescendo the vulgar creatures defecated in their hands and threw the fresh 'stuff' at de Loys and his men. Merde is all a Frenchman will take. De Loys cocked his rifle as the angry male stepped forward. At the last moment, it stepped aside and the snarling female advanced beyond the male, and there de Loys' bullet dropped her after a loud burst. "The other one, unfortunately wounded, managed to escape and disappeared into the jungle, the great thickness of which prevented its recovery," wrote de Loys in his own account in the *Illustrated London News*.

Although a geologist by training, de Loys realized he had a significant zoological find before him. He brought the female corpse back to his camp, propped her on a fuel crate and bracing a stick under the chin to hold her face up he took a perfect picture.

De Loys' picture and report took the zoological community by surprise. No one had seen an 'ape' version of a known genus of monkey before. The standing theory of evolution could not even accommodate such a concept. According to Gradualism, ape characteristics, which are closer to mankind, mark them out as higher primates. Bipedalism takes the species even further along the evolutionary ladder. It indicates something even closer to mankind than the great apes of Africa. Thus if de Loys' 'ape' was real it meant the discovery of the most advanced anthropoid next to Man. Yet its features showed it was related more closely to a monkey, some of the most primitive of primates.

Adding to the controversy, de Loys' description of their behavior also undeniably fit those of spider monkeys. They, too, will build themselves into such a fever pitch that at its crescendo they will defecate in their hands and throw the uninviting confection at people. A "monkey man" with a monkey mentality was just not acceptable. Therefore those who at first believed de Loys' picture thought that perhaps those ghastly "its" were only a pair of giant spider monkeys but not genuine tailless anthropoids.

Yet even this backhanded acceptance suggested de Loys was stupid or lying, for he had remained adamant that they were tailless and bipedal. A picture didn't seem worth a thousand words either, at least not in his favor. Only one photo survived the jungle trek and it only shows the anthropoid from the front. He was thus rancorously denounced, especially by Sir Arthur Keith in *Man*, as perpetrating a hoax. Keith believed he had merely propped on a crate nothing more than a regular spider monkey and hid its tail from the camera. The only real advantage Keith's arguments had was the then-dominant theory of gradualistic evolution. No one was expecting to see a "missing link" that had nothing to do with mankind. Worse for de Loys was the fact he did not bring any part of it out for study. Thus, despite the illogic of Keith's arguments, it seemed easy for many to believe his denouncement that a geologist was perpetrating a hoax in the remote wilderness.

Petroleum was the object of the search, it must be reiterated. De Loys was not in a position to preserve or bring out the specimen. Twenty men went into the jungle; only 4 emaciated men came out. De Loys obviously had no time or knowledge to perpetuate a zoological hoax. Thus some zoologists believed him.

Professor Georges Montandon was one of them. He was the first and staunchest supporter of de Loys and the man who had caused all the outburst of skepticism in 1929 when he finally introduced the "freak" to the zoological community as *Ameranthropoides loysi*— 'Mr. Loys' Ape-Like American.' After a careful study of de Loys' report and picture, Montandon was sure de Loys shot not a freak at all but a specimen of a real species. Many zoologists were shocked. Respected Montandon was actually declaring that there was an anthropoid 'ape' native to the New World, and he even went so far as to name it after de Loys.

Scoffing and acrimony aside, Montandon's classification was on sure ground. Whether zoologists liked it or not, de Loys' picture also showed the corpse had undeniable differences from a regular spider monkey. It had a more developed, rounder head, a more human expression, a flatter thorax, broader shoulders, more massive body and thicker arms— attributes of anthropoids, and in this case one taller than a gorilla.

The crate upon which the 'monkey' was set secured the height without a doubt. Montandon had gone to the trouble of getting a copy of the crate from his cousin in the oil business. The crate was 17 and 3/4 inches in height. Measurements from the picture showed the "monkey" was at least 3 and a 1/3 times taller than the crate, so that the height could have been as tall as 5 feet.

But the most amazing thing was that de Loys' picture confirmed that the 'ape' was indeed related to spider monkeys. For example, the female had enlarged sexual organs to the point that to a casual observer she would look like a male. Most notably, the eye sockets were surrounded by a boney brow or ridge, as in the jutting brow of apes but in this case it continued around encircling the eye. This more than anything is a trait identical to spider monkeys. A more subtle relation was uncovered in de Loys' report concerning its

*Copied from the picture taken by de Loys in 1920.*

teeth. All New World monkeys have 36 teeth whereas Old World monkeys have 32. Of all monkeys in the New World there is only one exception to the rule: the spider monkey; it sometimes has 32 teeth. De Loys had reported that the 'freak' had 32 teeth.

The *loysi*'s nose proved it was native American. All Old World primates are Catarrhinian, that is, having downward venting noses. All New World primates are Platyrrhinian, that is, have splayed nostrils divided by a thick wall of cartilage. It was clear from de Loys' picture that the *loysi* was Platyrrhinian. It definitely originated in America, not Asia or Africa.

Professor Georges Montandon's work on *Ameranthropoides loysi* went far beyond examination of the photograph, and his pronouncement of a new species was not hasty. Beleaguered by attacks in 1929, Montandon fought back in the wisest way: with knowledge. Confronted by the new discovery, Montandon had done extensive research and discovered that de Loys was only the last in an impressive line of people who had either encountered such a primate or had heard of them. Montandon first uncovered Pedro de Cieza de Leon's reports of South America. Within them there is found a description of the *loyses* as human monsters, along with an amusing pedigree to explain them.

> It also said that in other places there are (though for myself have not seen them) large long-tailed monkeys which live in the trees and which the natives (tempted by the devil who seeks to find where and how he can make men commit the vilest sins) use like women and, it is affirmed, some of these monkeys are supposed to give birth to monsters which have a man's head and privy parts and a monkey's hands and feet. They have, it is said, a thin body, and a great stature. They are hairy. Indeed they resemble (if it is true they exist) the devil their father. It is also said that they have no language but a plaintive moan or howl.

Tales and reports of such a foul "manbeast" were continually heard by the first European explorers. In 1595 and 1596 while exploring his new discovery of Guiana, Sir Walter Raleigh also heard from natives (through his liaison Lawrence Keymis) that there existed strange monsters in the jungle that resembled the "pans, fauns and satyrs" of Greek myth. Considering the size of the, shall we say, erogenous organs on the female that de Loys killed, all *Ameranthropoides loysi* would strike observers as males, which fits the Greek idea of the satyr, a race of bestial men with long exaggerated reproductive organs.

In 1769, Dr. Edward Bancroft also made the report of an "orangutan" in Guiana bigger than the "African" (chimp) and the "Oriental" (modern orangutan). His description is not at odds with *Ameranthropoides loysi* although he thought the Indians exaggerated its height when they said 5 feet tall because they "greatly dread them." They maintain "an erect position, and having a human form, thinly covered with short black hair."

In 1860, from the evidence he could obtain, naturalist Philip H. Gosse deduced and declared that in South America "there may exist a large anthropoid ape, not yet recognized by zoologists."

Further testimony uncovered the wide territory over which such 'apes' could migrate over South America. The Marquis de Wavrin lived for a long time among the Marakshito pygmy Indians in the Amazon. He was told that large monkeys are reported by the natives to exist, especially around the Orinoco. While he never saw any of these *maribundas*, they are supposed to be around 5 feet tall, walk upright, and their cry is "strangely like a human call." Monsieur le Marquis goes on: "On the Guaviare in particular I several times thought at first that it was Indians calling."

Among the historical accounts that prove startlingly corroborative of *Ameranthropoides loysi* is that of Charles Barrington Brown, the government surveyor of British Guiana. In 1868 he heard rumors that "hairy men" lived in the upper Mazaruni at the Venezuelan frontier. He inquired of these rumors because he had heard "a strange plaintive moan or howl" (just as de Leon describes). He also recorded something else: "The first night after leaving Paimah we heard a long, loud, and most melancholy whistle, proceeding from the direction of the depths of the forest, at which some of the men exclaimed, in an awed tone of voice, the 'Didi.' "

Brown went on to inquire. "The 'Didi' is said by the Indians to be a short, thick set, and powerful wild man, whose body is covered with hair, who lives in the forest. A belief in the existence of this fabulous creature is universal over the whole of British, Venezuelan and Brazilian Guiana." Brown later met a "half breed woodcutter" who had been attacked by a male and female Didi "in which he successfully resisted their attacks with an axe. In the fray he stated he was a good deal scratched."

Among the Indian descriptions of the "Didi" this curious characteristic or, perhaps better put, social behavior is found. It is said that these wild men are only to be seen in pairs, a male and female. Accounts of them all over the jungles of South America constantly reaffirm this and refer to their loud whistle and howling moan.

As further proof of their existence and their unique ability to migrate long distances, Dr. Montandon discovered that the Maya had statues of huge gorilla-like creatures. Though they now stand on the stumps of their thighs, for their legs are gone, they are over 5 feet high. "They have a strikingly apelike position. They have pronounced eyebrows, broad chests and a bent back." It is also readily apparent that they have anatomical similarities to de Loys' anthropoids from South America. To the uninformed observer they at first strike one as bisexual because one with apparent male organs carries or is suckling a baby— the sexual attributes of female spider monkeys which was also found on the giant female *Ameranthropoides loysi*. He continues: "coming from a gorilla-less country of the Maya, they are one of the unexplained curiosities of the historical museum of Merida

Furthermore, Dr. Montandon notes that there is no legend to explain these apelike statues. They have simply stood on a hill for as long as any can remember. Should we suppose that these strange anthropoids made an impression on the Maya as they migrated northward through their land? Because of its longer legs, it would not be difficult for this agile anthropoid to continue northward along the Sierra Madres through Mexico and up into California and from there further north. Always traveling in a pair male and female would also mean new populations could arise in the new territories where these anthropoids settled.

It may seem that I am merely making a point that a native American anthropoid existed in South America and potentially could have come northward. Proving *Ameranthropoides loysi*, however, is a little more to the point than that. One, more than those tantalizing reports of the Old World Almas, de Loys' photo shows us that an entirely separate type of primate, one closer to monkeys and yet more advanced than any ape, does exist. Moreover, Bigfooters have grasped at Indian artwork to show us that ape features do exist and therefore these must represent the Yeti-Gigantopithecus of the Old World. In order to finally sort this out, it is best at this juncture to merely continue with our trail northward. Taking up where Montandon left off, the trail of the *loysi* can indeed be shown to have led north to the Pacific Northwest. And this trail begins to prove more than a native American "ape." It begins a journey that uncovers the origin of much of Indian artwork.

An interesting report appeared in 1870 in the *Antioch Ledger*, of Antioch, California, while the incident was still fresh in the hunter's mind. The hunter's name is not mentioned, as the *Ledger* took it from another paper, but the name in other non-original sources is sometimes given as de Groot. So let's elect the same name here. This is perhaps the first detailed and believable encounter with an unknown American anthropoid before de Loys' 1920 encounter. It happened in the fall of 1869 near Orestimba Creek in Stanislaus County, California. De Groot wrote it in response to the paper's skeptical reaction to reports that a "gorilla" had been seen in Crow Canyon and afterwards on the hills thereabouts. It is a vivid description that deserves to be quoted in full:

Last fall I was hunting in the mountains about 20 miles south of here, and camped five or six days in one place, as I have done every season for the past 15 years. Several times I returned to camp, after a hunt, and saw that the ashes and charred sticks from the fireplace had been scattered about. An old hunter notices such things, and very soon gets curious to know the cause. Although my bedding and traps and little stores were not disturbed, as I could see, I was anxious to learn who or what it was that so regularly visited my camp, for clearly the half burnt sticks and cinders could not scatter themselves about.

I saw no tracks near the camp, as the hard ground covered with leaves would show none. So I started in the circle around the place, and 300 yards off, in damp sand, I struck the tracks of a man's feet, as I supposed, bare and of immense size. Now I was curious, sure, and I resolved to lay for the barefooted visitor. I accordingly took a position on the hillside, about 60 or 70 feet from the fire, and, securely hid in the brush, I waited and watched. Two hours and more I sat there and wondered if the owner of the feet would come again, and whether he imagined what an interest he had created in my inquiring mind, and finally what possessed him to be prowling about there with no shoes on.

The fireplace was on my right, and the spot where I saw the track was on my left, hid by the bushes. It was in this direction that my attention was mostly directed, thinking the visitor would appear there, besides, it was easier to sit and face that way. Suddenly I was surprised by a shrill whistle, such as boys produced with two fingers under their tongues, and turning quickly, I ejaculated: 'Good God!' as I saw the object of my solicitude standing beside the fire and looking suspiciously around. It was the image of a man, but could not have been human.

I was never so benumbed with astonishment before. The creature, whatever it was, stood fully 5 feet high, and disproportionately broad and square at the fore shoulders, with arms of great length. The legs were very short and the body long. The head was small compared with the rest of the creature, and appeared to be set upon the shoulders without a neck. The whole was covered with dark brown and cinnamon colored hair, quite long on some parts, that on the head standing in a shock and growing close down to the eyes, like a Digger Indian's.

As I looked he threw his head back and whistled again, and then stopped and grabbed a stick from the fire. This he swung round, until the fire at the end had gone out, when he repeated the maneuver. I was dumb, almost, and could only look. Fifteen minutes I sat and watched him as he whistled and scattered my fire about. I could easily have put a bullet through his head, but why should I kill him? Having amused himself, apparently, as he desired, with my fire, he started to go, and, having gone a short distance returned, and was joined by another-- a female, unmistakably when both turned and walked past me within 20 yards of where I sat, and disappeared in the brush.

I could not have had a better opportunity for observing them, as they were un-conscious of my presence. Their only object in visiting my camp seemed to be to amuse themselves with swinging lighted sticks around. I have told this story many times since then, and it has often raised an incredulous smile; but I have met one person who has seen the mysterious creatures, and a dozen of whom have come across their tracks at various places between here and Pacheco Pass.

While this report has easily resided in that convoluted world of Bigfoot dossiers, it resembles anything but that bloated legend. De Groot's account simply has too much similarity with those of South American Indians and European officials (5 feet tall, whistles, moans, howls, travels in male-female pairs) not to support the idea of the *loysi*'s genuine existence and northward migration.

If continuing a northward migration, there is also no reason why these two *Ameranthropoides loysi* would not follow the Sierras and along the Cascades into Oregon and into Washington and then British Columbia, following a migratory route that some of their own kind might have blazed in like manner thousands of years ago. It is when reaching the Pacific Northwest and the heart of Skoocoom and Sasquatch country that one realizes that de Groot's encounter and description was hardly an isolated event.

The artwork of the Indians of the Pacific Northwest proves that the *loysi* did indeed come northward. Othniel Marsh's discoveries of Indian stone heads in the Columbia Valley representing what immediately strikes one as ape heads has long created a conundrum inasmuch as they establish that Indians of the Pacific Northwest had knowledge of ape physiology, something that was not thought to be possible considering the firm belief that apes only existed in the Old World. For those who could not deny the immediate ape-like appearance, the only solution seemed to be the old theory that some apes had in ancient time migrated over the Bering Land Bridge like many other animals had done.

Examination of many of these heads, however, in light of estab-lished physiologic characteristics of the Ameranthropoid, reveal fea-tures unmistakably pointing to a purely South American origin and thus represent something native to America and not some imagi-nary bogey man of Indian myth or Old World ape. In a number of noteworthy heads a Platyrrhinian nose is obvious, and in one

Stone head showing Platyrrhinian nose and an ocular boney ridge.

Stone head showing Platyrrhinian nose.

Some young spider monkeys' hair grow in stocks "like a Digger Indian" until it is long enough to fold down. It gives the false impression of a cone head or sagittal crest.

Mature Spider Monkey.

Old female with hair loss and long hair on head.

Human expression in some spider monkeys.

rock head the ocular circumference existent both in *loysi* and spider monkeys is rendered with commendable accuracy and detail.

Evidence goes beyond weathered stone heads. Wood totems of the Dsonoqua are a case in point. More than any Indian legend, Bigfooters have promoted Dsonoqua as tangible support of Sasquatch-as-a-Yeti reality in America. Yet this Indian legend of a hairy nude people or evil spirits of the deep forest is represented in artwork as something far from a giant cone-headed rogue *Gigantopithecus*. Totems all contain the features of a heavy brow and ridge around the cheek, suggestive of *loysi*'s boney ocular ridge. Artistry has rendered the nose indistinguishable as a Platyrrhinian, but one other constant feature is remarkable. Dsonoqua is always rendered with the lips pursed as if howling or whistling. Other fea-

Dsonoqua totem head, Vancouver.

tures are protruding ears, large drooping breasts and nudity, attributes and lifestyles that do not reflect the Indians at all but those of the cannibal nude tribe that they insist inhabited the deep forests. Many times Dsonoqua is also portrayed as holding her child, an interesting corollary with those enigmatic Mayan statues thousands of miles away.

Tribal dance masks of various Indian cultures are also meant to represent the Dsonoqua. Some appear very human but nevertheless retaining the howling, puckered mouth. Yet others show features far more accurate with the *loysi*. Principally, the prominent

South American howler

brow circles around the eye and carries on around to the cheekbone. In one mask, a Platyrrhinian nose is also apparent by how a deep cleft separates the nostrils.

More than anything the full body ritual suit of the Koskimo Indians confirms for the toughest skeptics that *Ameranthropoides loysi* is real. It is said to represent the "Hami"— "Dangerous Thing." It is made of pieces of fur pelts that reproduce the entire body proportions of a bipedal anthropoid with huge hands. By the pursed, howl-

[ 143 ]

ing mouth the wooden mask that accompanies this outfit clearly represents a male Dsonoqua. The ocular area of the mask, however, confirms the Dsonoqua are the *loysi*— the unique ocular boney ridge, exclusive to the spider monkey and the *Ameranthropoides loysi,* is perfectly reproduced on the mask.

It is so close that Dr. Leonce Joleaud's study of *Ameranthropoides loysi* sounds like an artistic critique of this area of the mask, the similarity being "in the general arrangement of the ocular region, where the eye-socket cavity is surrounded by a projection of bone which is almost equally prominent everywhere and is carried on just as markedly across the cheekbones as in the eyebrows."

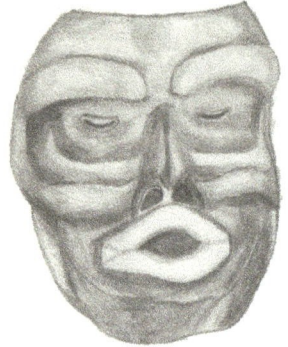

A picture of a Koskimo Indian wearing the full ritual outfit can be found in the Edward S. Curtis Collection, now in the custodianship of the National Archives. This collection contains the pictures he used in his 1914 book *The North American Indian*, 6 years before de Loys would shoot a 5 foot tall living version of just such a "Hami."

Dsonoqua Mask

In a strange and belated but profound way these representations of "cannibal peoples" by the Pacific Northwest Indians exonerate a geologist who was venomously denounced by many zoologists, even long after his death, because his picture remained as a testimony to something they did not suspect: a primate neither ape, man nor monkey. What he shot in South America clearly came north. Ironically, the evidence for a "Mr. Loys' Ape-like American" was always present but preserved best thousands of miles from where de Loys shot one. Clearly, because of its human-like proportions, the agile anthropoid was capable of impressive migrations.

It is equally possible that some could travel along the Sierra Madres through New Mexico, Texas and eventually along the Rockies until also arriving in Canada or branching off into other parts of the United States.

Modern migrations along this proposed route might explain the Arcadia Valley sighting of 1869 in Crawford County, Kansas. Although it is far away from the Pacific Northwest, Kansas is at the foot of the Rockies. Crawford County is not, but it does border the wild Indian Territory that later became Oklahoma, where one might have found wooded grounds to its liking. In 1869 a "savage"

Hami, of the Koskimo Indians.

terrorized the whole area. Mr. M.S. Trimble finally wrote a letter to the St. Louis, Missouri, *Democrat* which they published. In it he writes:

> It has so near a resemblance to the human form that the men are unwilling to shoot it. It is difficult to give a description of this wild man or animal. It has a stooping gait, very long arms with immense hands or claws; it has a hairy face and those who have been near it describe it as having a most ferocious expression of countenance; generally walks on its hind legs but sometimes on all fours. The beast or 'what is it?' is as cowardly as it is ugly and is next to impossible to get near enough to obtain a good view of it.

Trimble expressed his hope that if any showman had lost a gorilla or orangutan, or if a menagerie back east in St. Louis is missing one, they will see his letter and try to come and catch the thing. He wrote, with a definite air of frustration, how 60 men went hunting for it but it nonetheless eluded them. It was gone for several days, to their relief, but then it came back "as savage as ever."

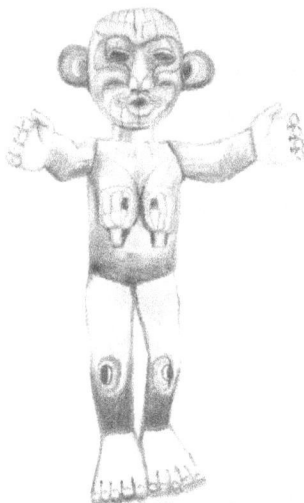

Trimble could almost be describing the Hami suit of the Koskimo Indians— the immense hands, long arms and human form— or the Dsonoqua standing totem— or de Loys' photo.

A description by a Saskahaua Indian of a Sasquatch seems to conform to the same mug shot. William Point's encounter occurred in 1927 and found mention in Burns' first *MacLean's* article and later was reproduced in his 1940 *Wide World* article.

Standing totem of Dsonoqua

Dear Mr. Burns,

I have your letters asking is it true or not that I saw a "wild giant" at Agassiz last September while with the hop-pickers there. It *is* true, and the facts are as follows: --

Adaline August and myself started for her parents' house, which is about four miles from the picnic grounds. We were walking on the railroad tracks when

Adaline noticed someone walking along the grade, coming toward us. I also saw this person, and at first thought it another man walking the tracks as we were. But as he came closer we noticed that his appearance was very strange, and on coming still closer we halted in amazement and alarm; we saw that the man wore no clothing at all and was covered with hair like an animal. We were both very frightened. I picked up two large stones with which I intended to hit him if he attempted to molest us, but within fifty feet or so he just stopped and looked at us.

He was twice as big as the average man, with arms so long that his hands almost touched the ground. His eyes were very large and as fierce as the cougar's. The lower part of his nose was wide and spread over the greater part of his face, which gave him a very repulsive appearance. Then my nerve failed me and I turned and ran. I looked back as I ran and saw that he had resumed his journey.

Adaline August had fled first, and she ran so fast that I did not overtake her until we reached the picnic grounds, where we told the story of our adventure. Older Indians who were present said that the monster we encountered was undoubtedly a "Sasquatch," a tribe of wild, hairy giants, now almost extinct, who live in the district in tunnels and caves.

Assuring you of the truth of this,

Yours truly,

William Point.

Unlike the many other descriptions Burns relates, this letter brings forward the picture of a strange primate, animal but human in appearance except for the much longer arms (it may have been stooping thus making the arms look longer). But there is that one curious tidbit about its eyes. In the shadows could Brave Point have mistaken the boney ridge around the eye as the entire eye and therefore thought it had huge eyes?

In looking at other classic reports we see more descriptions of the *loysi*. The famous 1906 incident in which Indians were so scared they paddled out to the steamer *Capilano* is another case in point. They described a hairy creature had come out to the beach to dig clams. It was 5 feet tall and started to howl, and, frankly, scared the hell out of them. In Alaska there were the reports of "monkey men" or the bushman who chased and in one instance killed a prospector.

The real apelike foot of the Ameranthropoid can also explain a lot of the old frontiersmen accounts of footprints. Spider monkeys, indeed probably most monkeys and apes, sometimes walk with their toes curled under. To an untrained eye this gives the impression of short, squat toes, especially for impressions in thick mud. When rushing or running, however, their feet are spread out flat. This would give the impression of long, spread toes, just as Mike King reported on Vancouver Island. Some spider monkeys have short "thumbs" or hold up their offset thumb/toe and thus leave a 4-toed print. Depending on what toes they are curling under, or whether they've lost a toe, the Ameranthropoid could leave a broad range of unusual footprints.

This sublimely reintroduces Bluff Creek into the equation. Months before Ray and Wilbur went on their escapades unusual tracks were found at Korbel Woods. They were so unusual that up to 25 people had stopped and followed them. All the hype of "Bigfoot" brought this incident belated coverage in the *Humboldt Times*, though buried amidst the hail of giant manbeast reports. On October 14, 1958, roving reporting Andrew Genzoli reports:

> Determined to trace the facts concerning the big feet, I called Julian Pawlus . . .who was among the men on the Korbel scene to witness the tracks. The tracks he had seen, he said, were observed last spring, while he, with others, were doing a logging road construction job. He described the tracks as being "pretty heavy" and made by "bird or human." He didn't wish to be carried away by what he had seen, and he referred to the mark as being of a "paw" rather than a foot.

He said the Korbel Woods impressions were found on the edge of a creek, and following the path, they went over a sixty foot culvert into a freshly graded road. The tracks were sunken enough to indicate the weight of the object. Pawlus said that there were three toes straight out and a couple of smaller on the side.

For the sake of the legend and for those who adore Bluff Creek's treasured chapter in it, I'm glad that such prints were recalled and reported. Unfortunately, however, the maker is not the berry-eating buddy of modern legend. Pawlus describes a foot that could very well belong to a *loysi*. If it curled under (or had lost) its first toe it would certainly look in combination with the "thumb" to be another small toe off to the side. At the very least this indicates that an anthropoid primate could have been moving through the area as late as 1958. Along with the *loysi* there comes all the nasty stuff attributed to the wild men in the old journals— probably more so if Brave Point was right about a much larger size to the specimen he saw.

*A spider monkey (and hence* Loysi*) could make all of these prints if the thumb didn't mark, if it was curling under one or all fingers or had lost a toe. Pawlus said the Korbel Print looked like it could have been made by a bird, which means it had a very narrow heel.*

Except for no clear report of the Dsonoqua or "wild man of the woods" flinging dung at Indians and prospectors, his behavior conforms to that of the *loysi* in South America.

True, a crap-flinging monkey does not create the attractive picture of the missing link or Gentle Giant *Gigantopithecus* the modern legend has painted, but it is really something far more believable than the legend and supported by several independent witnesses over centuries and by the tangible Indian artwork.

The truth behind the Indian histories of a violent people of the forests can also be far more appreciated considering the *loysi*'s origins amongst the monkeys. Old World apes can be largely altruistic to people, but monkeys are often violent. Monkeys not raised around humans can and will often go into fits when first seeing humans. Monkeys can be omnivorous, violent, and loud. Ape behavior ranges from orangutans making lip smacks and grunts; chimps stupid shrieks like a spoiled child; and gorillas, the introverted mammoths of the apes, seldom unleash their retching cries. On the other hand, certain species of monkeys can sound like wolves, howling and moaning to an unnerving pitch. For instance, the South American Howler is the second loudest mammal on Earth after the Blue Whale.

Far more than apes, monkeys will band together into violent "tribes." Recently this was demonstrated with unique tragedy when on October 23, 2007, Macacas monkeys ganged together and attacked S.S. Bajwa, the Deputy Mayor of New Delhi, India. In struggling to fend off the marauding monkeys he fell from his home's balcony and plummeted to the ground, sustaining head injuries that took his life in the hospital the next day. The Macacas, interestingly enough, have an unusual resemblance to the Almas, each described of the same color and having pointed years.

Perhaps a little more on the humorous side, hordes of spider monkeys invaded Playa Hermosa, Costa Rica, in September 2006 and congested and all around disturbed the surfing Mecca. Queuing up, surfers were pelted with dung. "Locals also claim that the monkeys blatantly and indiscriminately drop in on their human counterparts, and heckle anyone paddling out, thus denigrating the surfing vibe."

Visual descriptions of "hairy hominids" vary to the point that they don't always inspire confidence. But behavior patterns have proven remarkably consistent in far-flung witness testimony. This is true not only in the Americas but in Eurasia. Sir John Hunt relayed the following incident to Ralph Izzard concerning the encounter had by Hunt's Sherpa guide Tenzing Norkey's father in the yak village of Macherma, an essentially out-of-the-way place in a steep valley where there were a few stone huts for the itinerant shepherds and some seasonal grazing areas for the yaks.

Tenzing senior had taken his yaks and goats to Macherma in order that they might profit by the summer grazing. One day he had taken his cattle down to the valley floor to pasture as usual when he suddenly saw them herd together in extreme alarm. He seized his staff fearing a wolf or leopard, but on looking around was astonished to see that the cause of the excitement was what seemed to him to be a small man who was bounding down the mountain-side towards the central stream. The creature was covered in reddish-brown hair and was about 5 feet tall. The skull was conical in shape and the hair of the head being especially long, it fell over the animal's eyes. The features were those of an ape but the mouth was especially wide showing prominent teeth. The creature walked on two legs like a man; he was convinced he was confronted by a Yeti and rounding up his yaks, who needed no encouragement, he drove them into one of the stone huts and barred the door. This did not deter the Yeti which sprang on the build ing and began to dislodge a portion of the roof (This would not be difficult for roofs in Macherma are constructed of shingles held down by boulders.)

Tenzing senior survived because he kept his wits and quickly lit a fire and added dried chilies to it. The fumes, rising through the opening roof, quickly drove the Yeti away. It "jumped to the ground and ran in circles tearing up small shrubs and uprooting rocks from the turf, chattering with rage." Tenzing senior fell serious ill afterward from the shock of the encounter and did not recover for over a year.

In the Himalayas in 1954, of course, nobody knew about Fred Beck's encounter with the "gorillas of Mount St. Helens" in 1924 and Sasquatch's (and Skoocoom's) propensity to band together and be violent. Yet the image that de Loys' discovery and Indian stories paint is remarkably the same as that image natives of Eurasia have painted for the Old World. A giant ogre is not the terror. It is a man-sized primate that doesn't mind going on the rage after a human, even to the point of trying to get into the house. The most significant thing about these corollaries is that the Ameranthropoid cannot be the Yeti and vice versa, nor are either of them the Almas. Descriptions are just too radically different. They are separate and distinct species but all apparently of a similar type, one that reflects the world of monkeys.

Instinctively, it would seem easy to dismiss the whole idea that there could be several, maybe even 3 or 4, species of undiscovered anthropoids about, but we must put the reports of such creatures in the context of their geography. Eurasia, the American Pacific Northwest, and the Amazon, are huge territories. Their ability to elude capture should not make us regard them as fairy tales. Rather, we must remember that the waddling mountain gorilla remained unproven until 1901. This being said it should not strike us as surprising that primates with long legs and great mobility should continue to evade us. Humans have done so, of course, even near heavily populated areas. Shoichi Yokoi made world news in 1972 when he was captured on Guam. For 28 years he had hidden on the small island, the last Japanese soldier there from World War II to surrender. Two years later Lt. Hiroo Onoda surrendered in the Philippines, he too having hid in the jungles since the closing days of the Second World War. What can the vastness of Eurasia and the Pacific Northwest conceal?

More than anything the descriptions of the Skoocoom force any researcher to consider more than one species is afoot. The miners' encounters on Mount St. Helens proved the Indian histories that the Skoocoom were mountain dwellers, not forest dweller like the Dsonoqua. It's also very hard to imagine that the *loysi* under any circumstances could make a footprint as large as that described at Bannock County in 1902 or at Mount St. Helens in 1924. Nor is it probable it could consistently make a 4-toed print.

Until 1973 there was room to doubt the genuine legitimacy of 4-toed prints. But the pliable ambiguity of old reports was done away with on September 29 in Manitoba when three 4-toed prints 21 inches long were found by the conservation officer, R.H. Ucht-mann. These clearly stand separate from anything a *loysi* can make. In his letter to the Manitoba Museum of Man he stated: "There were imprints of four large rounded toes almost in a straight line across the front. . .The print is generally flat from heel to toes." Uchtmann also recorded that the prints sank in about 1 and 1/4 inches into the soil whereas his boot only marked 1/4 inch. A few days later Uchtmann led a wildlife biologist, R.J. Robertson, and a wildlife technician, B.E. Jahn, back to the prints to examine them. The photo he took shows us more than anything that something large comes from the land of the monkeys. It is remarkably mon-key-like, long and narrow. But from there it is incredibly different, more human and even more like the Ruby Creek Print.

Although this sighting took place in Manitoba, this foot can be identified with the Skoocoom of the Pacific Northwest. British Co-lumbia, the very heart of the legendary country, produced a very similar pair in 1980.

The incident and sighting happened at Dunn Lake near Berrière in April. While Tim Meissner was fishing with a friend they heard a loud mewing screech and looked over. There was a giant. "He was about 9 feet tall, black and hairy. He had a human-like face. . ." It promptly ran off into the woods. Meissner and his friend went over to investigate, there finding a dead deer which had had its neck broken. Three days later Meissner saw the creature again and was so scared he instinctively shot at it, but missed.

Although I personally am skeptical of some of the story, especial-ly the 9 foot height (as in the Sixes River monster, 9 feet was re-duced to 7 feet by the next witnesses who saw it), there is a reason to believe some of the reports around Dunn Lake that month. With-in a few days agricultural teacher Jack Wood found huge 4-toed footprints 16 inches long and 10 inches wide (suffering from distor-tion). It is wide at the ball and tapers to the heel.

This type of footprint was by this time not even associated with

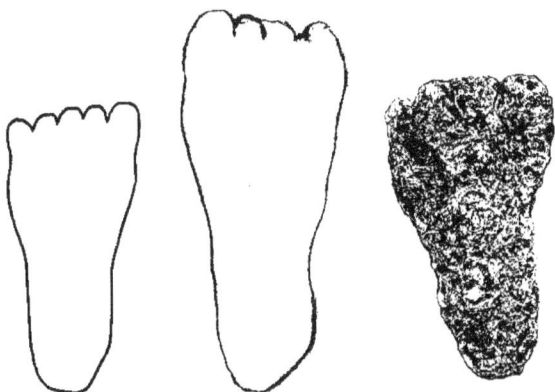

*The Ruby Creek tracing compared to the tracing of The Pas print found in the Canadian Shield area of northern Manitoba, 1973. The heel is nearly identical. Distorted but intact, the Dunn Lake Print, 1980.*

anything remotely Sasquatch or Bigfoot. This was the era of enlarged human feet and missing links. The 4-toed prints were but mere freaks that could be dismissed. Yet those who researched these incidents should have realized they were looking at the feet of the legendary Skoocoom. We're dealing again with that damn giant "human monster" of Bannock County and that "race of beings of a different species who are cannibals" on Mount St. Helens.

In 1988 this was reinforced again by yet another incredible find. The source is no less than the Royal Canadian Mounted Police. They were about 200 miles north of Winnipeg, north of Lake Winnipeg, when they came across and photographed (and removed) a set of tracks that crossed a dirt road. These are unquestionably 4-toed tracks with narrow heels. Thanks to the Canadian law enforcement, there is excellent evidence that the giant "human monster" of Bannock County, Idaho, and the Skoocooms are still living in Canada.

But how to classify the Skoocoom?

More trouble ensues in trying to do so, for they can't be the Sasquatch. One "tribe" of Sasquatches could speak and the other must be the owner of that 5-toed Ruby Creek Print. Yet something must own those 4-toed prints. Instead of gazing across the Bering Straits, it might be best to consider our own backyard again. Just as the Old World may have more than one species of anthropoids, perhaps so has the New World. Amongst *Ameranthropoides loysi*, is there another species of South American anthropoids akin to another species of monkeys?

Robert W. Morgan uncovered a valuable clue for "Sasquatch."

[ 153 ]

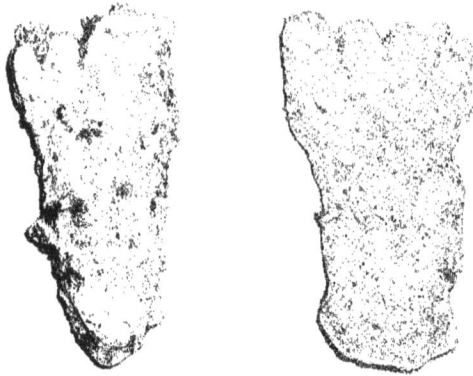

*Actual bitmaps of the prints, as contained in the Royal Canadian Mounted police photograph, taken in 1988 in Manitoba. The print at left (which would be of the right foot) has the larger toes obscured. These are classic examples of narrow heels and "spreading toes."*

His documentary *In Search of Bigfoot* (1975) made the point that "Sasquatch" had 5 fingers and no thumb, that is, its thumb was jointed like its fingers and was not opposable as in man and apes. This is not a trait, of course, of *Ameranthropoides loysi.* It has 4 digits and not 5, just like its monkey cousin. But this trait is significant for being the trait of the South American Howler monkey, the one New World monkey that actually makes a sound approximating the eerie wail for which the "Sasquatch" is also noted. What is more significant about Morgan's discovery is its location. Morgan concentrated his research around Mount St. Helens, so it seems he was actually being told of *Skoocoom* traits. It is only our popular but innocent mistake that Skoocoom and Sasquatch are one and the same.

This is not the only attribute of Howler Monkeys that has been reported in the "Skoocoom." Fred Beck and the miners reported that the ears of the Skoocoom stood out and were easily definable— a trait of Howlers. Also, the miners' attackers were narrow in the hips and broad in the shoulders, also a trait of Howlers. Also again, the Skoocoom had beards. Paul Kane most certainly must have been told of this trait. When he encountered Indians watching him warily from the brush along the Columbia River, he wrote "As I sat upon the packs taken from the horse, nodding in silence, with fixed stare at them whichever way they turned, my double barrel gun cocked, across my knees, and a large red beard (an object of great wonder to all Indians) hanging half way down my breast, I was, no doubt, a very good embodiment of their idea of a manitoum, or evil

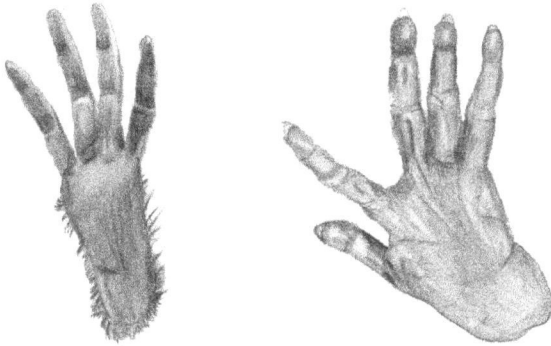

*Spider monkey's long 4 digit hand. Howler monkey hand. Its thumb is jointed like a finger. It is not opposable like a human thumb.*

genius."

This is an extremely noticeable trait of Howlers, which have a thick, almost Santa Claus beard.

Most powerful again is that wraith, Coincidence. Long before Beck, there was an obscure report that described a "wild man" along the Cascades. The report is dated 1886 from Happy Camp, California, in Humboldt/Del Norte Counties, the heart of Bigfoot country. In the *Del Norte Record*, dated January 2, we find written:

> I cannot remember to have seen any reference to the "Wild Man" which haunts this part of the country, so I shall allude to him briefly. Not a great while since, Mr. Jack Dover, one of our most trustworthy citizens, while hunting saw an object standing 150 yards from him picking berries or tender shoots from the bushes. The thing was of gigantic size— about 7 feet high— with a bulldog head, short ears and long hair; it was also furnished with a beard, and was free from hair on such parts of the body as is common among men. Its voice was shrill, or soprano, and very human, like that of a woman in great fear. Mr. Dover could not see its footprints as it walked on hard soil. He aimed his gun at the animal, or whatever it is, several times, but because it was so human would not shoot. The range of the curiosity is between Marble Mountain and the vicinity of Happy Camp. A number of people have seen it and all agree in their descriptions except some make it taller than others. It is apparently herbivorous and makes winter quarters in some of the caves of Marble Mountain.

A "bulldog's head" is another most outstanding feature on the Howler. It does not have a cone head or even a rounded one. It has a low vaulted head with that familiar bulldog cleft down the center.

Despite the fraud of Bigfoot, the wooden-footed gadfly walks the land of the very real legends of the Omah, the hairy manbeasts of Indian legend. It is because of these Indian histories that John Green and so many others believed Bigfoot walked a real pedigree. Jerry Crew and Betty Allen even talked to the old Indians, uncovering the Omahs were noted for their "eerie howls."

The Omah and Skoocoom appear to be the same thing. But they do not walk on enlarged human feet. They walk on long, monkey-like feet with 4 toes. They are, in essence, the true Bigfoot. Like the Dsonoqua (*Ameranthropoides loysi*), they are purely ours, purely American. It was our— White Man's— mistake to equate Dsonoqua with Bigfoot and make it evidence for a giant Eurasian Yeti. Rather they are native American anthropoids, with the Omah and Skoocoom being something more along the lines of the legend. They are close to 7 feet, which is truly huge. That is the height of the door frame of an average modern home. Imagine a hairy primate whose head is close to touching the top of the frame and whose thick body fills the doorway. That is enough to rate a giant by anybody's standard.

The habitat of the Skoo-coom also suggests it comes from South America. It has been reported along the Cas-cades, Klamaths and into Washington State, Mount St. Helens being famous as its

Face of the Skoocoom?

remaining stronghold. This does not indicate a primate that has migrated to America from Eurasia. One would expect more reports in Alaska if that was the case.

This supposition is made stronger by returning to South America to take up the trail again. In a disturbing coincidence we also find accounts of something bigger that perhaps should also be taken as equally authentic as *loysi* has been in Indian artwork. No less than Alexander von Humboldt reported in 1799 that when he reached the rapids of the Orinoco River (first white man to do so) he heard of a giant "hairy wild man of the woods" that supposedly built huts, carried off women and ate human flesh. He discovered that even the missionarion in the neighboring countryside believed in "va-

[ 156 ]

sitri," as the natives called it— "big devil."

In recounting the "man faced monsters that haunted the jungles of Brasil" in *Noticias Curiosas e Necesarias das Cosas do Brasil* the Jesuit father Simão de Vasconcellas, S.J. provincial of Brazil, speaks of something the good padre could only call a giant ogre. He records it to reach as high as 11 feet tall. It is considered the greatest and most courageous of all the hairy wild men. It is hard to determine, however, if it is a man or beast. It is certainly regarded as between both. The tribe is called the *curinquéans*.

This name is amazingly phonetically close to the Pacific Northwest Indian pronunciation of Karankawa or Karakawas. When Indians would talk with Whites about the origin of the Sasquatch men they would use this name. Both Haisla and Salish Indians claim that Saskahaua Georges came from the south along the mountains, driven thereto to take refuge from the Spanish conquistadores. There they degenerated to become dumb, hairy cannibalistic brutes. Hearing them called the Karakawas or Karankawas, White Man interpreted the Indians to mean the Texas tribe of Karakawas. However, in light of Vasconcellas' *Curiosas*, it is tempting to consider the Indians were referring to these same *curinquéans*.

Legends of giants abound in South America, the *curinquéans* being more obscure and one applying to central and northern territories of the continent. In the south, the "Giants of Patagonia" (south Argentina/Chile) became quite a famous controversy. European explorers and traders in the 16th and 17th centuries insisted such a people did exist and were 11 feet tall. In reality, they averaged 6.6 with the largest 7.6 feet tall (still large!). If the "Karakawa" and the *curinquéans* are one and the same thing, no doubt the same can be said of them as for the Patagonians. It is more than likely they are closer to the height to 6.6 to 7.6 feet on the outside, just like the Indians reported the Skoocoom height to be.

If I am right, until now the actual Bigfoot has been faceless. He has been nothing but a comical legend that has done away with the truth, a part of which spoke of the unusual dread these creatures inspire in us and even in other animals.

Snippets of the old legend of the Skoocoom were still intact in 1950, eight years before "Bigfoot" would change it all. A disappearance that year brought to the surface the old "gorillas" of Mount St. Helens. This was the case of the disappearance of Jim Carter from a skiing party up the taboo mountain in May of that year. During the party's descent down the slopes, Carter stopped at Dog's Head at the 8,000 foot mark, where he asked the others to continue descending so he could photograph them. After this he was never seen again.

The next morning a search party found an empty film box where Carter had reloaded. The only other evidence were his ski tracks heading downhill, revealing to the searchers that he must have been "taking chances that no skier of his caliber would take unless something was terribly wrong or he was being pursued."

The article goes on to say that out of the 75 searchers there were many who had uncomfortable feelings during the search, as if a 'presence' was watching them while they were in the area. One, Bob Lee, was quoted in the *Oregon Journal*: "It was the most eerie experience I've ever had. I could feel the hair on the back of my neck standing up. . . .I was unarmed except for my ice axe and, believe me, I never let go of that." Lee and the surgeon for the Seattle Mountain Rescue Council, Dr. Otto Trott, were allowed to express their opinions that "the apes got him."

While I personally have my doubts about the Carter incident (the way it is presented anyway, since there are no reports of footprints), there is no doubt about the persistence of the legend of the "race of beings of a different species" on Mount St. Helens and the fear they inspired in the local Indians.

Ralph Izzard's initial investigation of the Nepal Himalayas, as an adjunct to his assignment, also left him with an unusual impression. "Many of us, and not only the hypersensitive, have felt the sense of a 'presence' at high altitudes. . .Nine times out of ten this 'presence' is felt to be malevolent. . ."

Rather than invoking psychic vibes or a sixth sense my guess is that these feelings are inspired by the pre-knowledge that one could encounter something like a Yeti or Skoocoom in these areas. Regardless of Izzard's designation for it as "Animal X," it is plain to detect that the Yeti held a particular aura of dread no other animal held. Throughout his writings it is clear that this unusual fear is caused by the fact the Yeti walks on hind legs like a man. There is something undeniably unnerving about being stalked by something walking like a man and yet only an animal. No four-footed creature has been able to instill the calculated predation that seems to come with a bipedal walk. Perhaps it is because we naturally associate bipedal with intelligence and thus from there it is a short leap to consider something capable of stalking as possessing calculated evil.

Just from seeing a clear track and print in 1951, Eric Shipton was profoundly affected. His account, published in his work *Menlung La*, reveals the eerie effect the encounter had on them as he reconsidered the day's event at twilight. "It was a clear, cold night," Shipton writes, "and as we settled down in our sleeping bags, the silence outside the tent was broken only by the occasional creaking and groaning caused by the movement of the glacier. I could not altogether suppress an eerie feeling at the thought that somewhere in that silent darkness, not so very far away, the strange creatures that had preceded us down the glacier were lurking. I was not surprised that Sen Tensing, who was lying beside me, was occupied with similar thoughts." Tensing sought to comfort Shipton by saying that the Yeti would be frightened since no humans have been here before. Whether Tensing believed it or not, Shipton writes: "I found this assurance comforting. . .But for the next few days I was very conscious of the presence of the monsters, and often wondered if they were watching our movements from some cave, of which we saw plenty in the vicinity."

The encounter clearly impressed upon Shipton the reality "of the existence of a large ape-like creature." He later declared: "Here we were in the presence of something truly unknown."

Perhaps one other thing inspires the instinctive dread in mankind— hands. An animal jumping on the roof and removing tiles to get at you just seems to excite the whole idea of an intelligent evil predator. Most any animal has the wits to do so, but it doesn't have the hands. In my own personal experience I've had raccoons rip a hole in my shake roof to try and get in. It provokes cussing from the homeowner (me, in this case) and a musing smile from the roof repair man who is grateful such critters exist and give him added business. But consider if those little b$#&%@$ had hands. They could simply pull off the shakes and jump in. Hands, it's hands more than anything that convey the sense of calculated intelligence and evil intent.

Add to this the unique proportions a monkey has as opposed to apes. Apes are short-legged waddlers. Many monkeys actually have very long hind legs. Now morph one into a 6-foot tall anthropoid, make it bipedal and remove the tail. We now have something far more human in proportion than any waddling ape. It could retain the small head of some monkeys and have long legs, just as Jeannie Chapman described the Sasquatch man at Ruby Creek.

Unquestionably, monkey attributes seem to be the one constant that run through these stories. The crest of the Yeti, for instance, is something also seen on the Crested Celebes Macaque, not apes. The Almas itself also fits this mold. In what is perhaps the latest sighting of one, in 1991 A. Tishkov reported that the creature he saw sported what looked like two tufts, feathers or shocks of hair standing upright above its ears. From photos that he took, Dr. Michael Trachtengerts drew sketches. This shows a hideous half-monkey, half-ape with a nose that curls down, almost hiding the nostrils. Yet Tishkov's description sounds suspiciously close to a known monkey— the Rhesus Macaque. Sometimes tufts of its hair stand up around its pointed, ever-so Almas-like ears— another mysterious "wild man" with a trait in common with a local monkey.

*An Almas, based on the Rhesus Monkey.*

Adding to this monkey image is the strange curve-down nose seen on some Amerindian totems and on the *bukwas*. In many respects this is the nose of the gibbon, the only Old World ape that is known to bark and howl and has, interestingly, monkey or bear-like eyes, those very eyes represented as beady spots on *bukwas* masks.

In some respects the gibbon strikes one as the *loysi* does. It's awfully monkey-like for an "ape." They bark, howl, and yet when on the ground (rare) walk habitually bipedal. Of all the apes, they can be quite aggressive, another monkey-like attribute. With its more human-like nose, they look like little hairy men, except for their very long arms. They really are a zoological oddity that we take for granted.

Lemurs are another oddity. They have a bear or dog's head and a monkey's body and hands. Platypus is yet another. Zoologically speaking, freaks really aren't that rare. They simply don't strike us as that enchanting because they are not mysteries. Like all other animals we have taken them for granted because we know they exist and we have catalogued them. Yet they, like so many other animals, reveal the abundant potential in nature.

Giant "versions" of known animals aren't rare either. The lion itself is a relative of the common domestic house cat. But the two are hardly alike. Larger "versions" also seem consistently to be more voracious and carnivorous. Stories of a man-eating cat in Africa similar to the common cheetah went unproven until 1975 when Paul and Lena Bottriel finally tracked and photographed the "King Cheetah" along the Mozambique border. This was undeniable proof that the common and relatively harmless cheetah has a rare giant cousin that in personality is radically different and dangerous.

*Gibbons, male and female. Hair growth gives the male the appearance of a cone-head. Facial hair growth also gives them a similar appearance to the* loysi *and spider monkeys.*

The repercussions are enormous if there really exist a type of large primate physically similar to man but with no affinity to us but to monkeys. It offers the only bipedal predator to man (other than man) with the ferocious attributes of some of the most primitive primates. It gives us a bigger version of the Macacas monkeys that clubbed together and eliminated the deputy mayor of New Delhi; and a more voracious species that could earn itself the reputation as a cannibalistic tribe on Mount St. Helens. They could do everything the old frontier stories describe. They could hurl stones at Caulfield Anderson and trash the miners' cabin in 1924.

However acceptable this conjecture might be to explain the origin of "Sasquatch," there is no denying that the Indians and then White Man did and do mix descriptions together so that the sum of the parts cannot possibly exist as popularly proffered. Sasquatch as we know it cannot exist and be a dozen different species— thin, thick, short, giant, animal, human— with contradictory attributes. The truth of the sum therefore lies in the parts when divided and sorted out. One part is divided and identified by understanding that *Ameranthropoides loysi* is a large piece of the puzzle. He is the odd anthropoid out. He is the one that can definitely be identified as the

*Loyses, illustrations based on spider monkeys, to which they seem related.*

Dsonoqua. Different tribes had various opinions about the Dsonoqua— from supernatural entities to simple "dangerous things"— but true people they were not. Indian artwork has always exposed Dsonoqua's attributes. Indians may have eventually confused descriptions, but White Man alone added Dsonoqua and Skoocoom as Sasquatch candidates. When Indians referred to two Sasquatch tribes they were *not* including Dsonoqua. When they spoke of the Skoocoom, they did in the context of something manlike but not man. When the Salish Indians, however, spoke of two tribes they meant *both tribes* were human.

From this point forward we have to divide between *these* two tribes. This is not to minimize the existence of the *loysi* and Skoocoom. They are a major part of the quest. They are that clue that allows us to divide between man, ape and anthropoids. The *loysi* has proven to be the clue that tells us a type of primate may be responsible, not an unknown species. But The Ruby Creek Print adds to the controversy and conundrum. That footprint is the most manlike footprint of any nonhuman species. It must be the true Sas-

quatch. It must be one of the two tribes to which the Indians referred. But it cannot be human. Yet it was so human in appearance
that the Indians believed it to be human. It has been confused for a
man on several occasions. And we have discovered through the Russian studies to respect Indian descriptions of humans, wild and primitive and yet very apelike.

The same trail that the Russians dared to walk could lead to
America— humans and a very human-like Old World primate. But
in order to narrow down our target, in order to weed out ape and
man, animal and troglodyte, we must follow a still older trail. How
can it be that there were (or still are) troglodytes or, for lack of a better concept, "primitive men"? How can such anthropoids so akin to
monkeys come about? We must follow an ancient trail, one with
clearer tracks than bogus cast in California plaster.

# CHAPTER 8

## Gradually Changing

By having just scratched the surface of historical reports and by having collated the accurate footprints we have uncovered a truly profound enigma, both in the Old and New World. This enigma hangs on Exhibits A & B, the Shipton Print and the Ruby Creek Print respectively. From there it expands to something similar— Exhibit C is ostensibly the 4-toed Skoocoom Print. Exhibit E we have the McNeely/Almas Print. Yet with Exhibit D it expands into something quite unexpected. The Gool ("Abnauayu") foot from the Caucasus is the one genuinely reconcilable print. Four of these prints are clearly not human, and the human print (D) is suggestive of Neanderthals, a very different kind of people we know once lived and whose origins anthropologists are still trying to decipher.

It is my contention that the evidence so far presented in this book points to two primary things. First, and most obvious, it indicates a kind of human being very different from ourselves in significant and consistent ways. Second, it indicates an unclassified *type* of primate. I won't say species because the circumstantial evidence indicates Yeti, Sasquatch, and Almas are of different families but of a similar

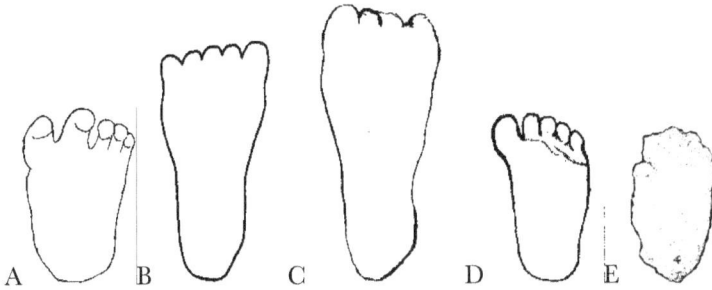

A    B    C    D    E

*Exhibits of evidence: the Yeti foot, as taken from the famous Shipton photo, 1951; the Sasquatch foot, as traced at Ruby Creek, 1941; A 4-toed print found in Manitoba and British Columbia, 1973, 1980; the Abnauayu footprint found by Dr. Jeanne Kofman in the Caucasus, 1981; McNeely, 1972, Nepal.*

type. From the footprint evidence in America we know that the Sasquatch and the Skoocoom must be different specie. But they all appear to be the same general type; and this type is a simian physiologically more like man than the apes but whose relationship is not to us but to monkeys. Siminids might be the best word— "simian-like;" or Simio-hominids.

How horribly erudite that sounds! But that is not my intention. Certainly if I desired to propose a new theory on primatology I would not choose such a topic to base it on. And that is not my world anyway. Mine is one of exploration and discovery, history, logic and philosophy. My intent here, truly, is to restore the original image of the quest. This image is of 1958; of the woodsman or dedicated hunter and tracker clad in the plaid Scotch flannel shirt, sporting thick-rimmed glasses and crowned with an Elmer Fudd hat. Defying scientific ridicule they went out into the dense forests of the Pacific Northwest on the tracks of big footprints. From there, unfortunately, they built a fantasized creature that roamed over the whole of America, if not the world. But the early image was a good image, a truer image and that, of course, makes it a better image.

But Science, like many other things, happens, and it should not be considered so dull and dreadful. It is as Einstein called it, "merely the refinement of everyday thinking." Any science begins with a collation of the original data. To do so is to uncover what we have uncovered. We have not uncovered a cone-headed giant. We have uncovered hints of something truly provocative. Therefore before we don that cotton flannel Scotch plaid shirt, sling our rifle over our shoulder and start tracking the deep unexplored forests, it would be negligent on our part if we did not try to explain, at least philosophically, where both an unusual kind of human and type of primate

[ 165 ]

could come from. Popular science, monstrously lagging behind the real thing, is still saturated by old gradualistic evolutionary theory and folk concepts.

Perpetual evolution is one. The other is a hierarchy mentality, that is, that everything will evolve to complexity and all biologic life can be placed in a neat ascending graph with Man at the top. In such a medium, "subhuman" and "missing link" are natural steps, despite being biologic absurdities. More than anything, it is my desire to banish these very old and fortunately antiquated notions. Only when free of these can we get a grasp as to the probable origins of "siminids" and extreme variations in human kind.

Some might not consider this a topic worthy of sufficient attention as to deem it necessary to update it with modern science and theories. Yet I would think it very serious indeed considering that the Soviet Snowman Commission's study into Zana has reopened the door on potentially explosive issues. She conjures up the old idea of ape men and missing links, that visual impression that Gradualism so beguiled science with throughout the greater part of the 20th century. It was because that image of early man was so predominant in the 1950s that Porshnev was sure that Zana could be nothing but a Neanderthal and therewith a "hominid relic," some representative of a past stage of evolution still living.

And indeed if ever old Gradualism had a poster child for the concept of an ape-man and mankind's evolution from a lower primate it was the skeletons of these very selfsame mysterious, cave-dwelling people. Visually their skeleton is entirely human, except their bones are thicker and denser than even the strongest weight lifter of today. But their strangely low-vaulted skull impressed observers as inhuman, with receding chin and jutting jaws that almost seem ape-like. Even more in-keeping with 'missing link' concept, the first skeleton that was found (in a cave in the Neander Valley of Germany) had a spine that showed the figure had a stooping posture.

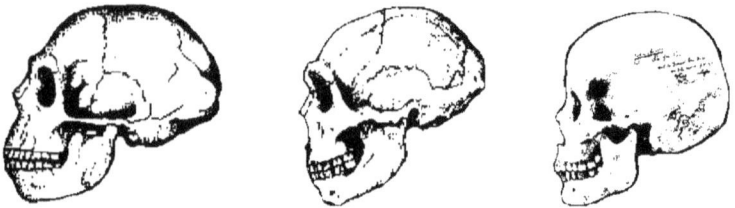

*"Classic" Neanderthal skulls compared to a modern human skull.*

[ 166 ]

From this we built a hairy, brutish ape-man, the true missing link between us and the apes. But time and more discoveries slowly began to change our impressions. Too many other skeletons were found that had normal posture. Further tests also proved that the Neander specimen suffered from severe osteoarthritis. After some 150 years of evolving views, today we picture all Neanderthals as a thickset people, between 5.6 to 6.5 feet tall, who could pass unnoticed in a crowd if properly manicured and dressed in appropriate clothes. Recent studies have even shown that some Neanderthals also possessed a gene MC1R indicating that some of them had red hair, freckles, and light skin.

Therefore the problem that Zana represents is very real. Her description is that of our 1950s' image of an "ape-man." But the footprints recovered from the area where she lived are those of Neanderthals. Whose image is true: the 1950s' Rice Boroughs' novelette or modern political correctness?

One thing is generally accepted by anthropologists: there were two "types" of Neanderthals— the "Classic" and the "Regulars." The "Classic" is the kind that was first found in caves across Europe. They had the low-vaulted, almost simian skull. The "Regulars" showed signs of both "normal humans" and the "Classics." Burials have proven that they wore clothes, had rituals for the dead, and even played music. Some anthropologists have offered a solution: that probably Neanderthals and our ancestors intermarried producing a mixed breed (the "Regulars") of stronger people showing the characteristics of both. This is a nice solution and probably true, but this also mutely admits that the "Classic" Neanderthal remains a very different type of human than 'us' and the "Regulars."

When in the 1970s we began to draw our new image, anthropologists highlighted how there was nothing about the Classic Neanderthal skeletons, naturally, that could speak as to how hairy these people had been. But the circumstances of their lifestyles should perhaps have suggested something unusual. The fact that the Classic was found in caves, appeared to have walked barefoot, wore no clothes, and could handle if not prefer cold weather, would indicate that they were very hirsute, did not care for warm weather, and were a "wild human." But, yet again, as we know now, that they could interbreed with us meant they were human.

This scenario is almost exactly, if not *exactly*, what Dr. Porshnev believed he experienced when tracing the descendants of Zana. Children and grandchildren had traits we do not have, all of them coming from that cave-dwelling grandmother. If Zana was indeed a representative of a living "Classic" Neanderthal, then our current imagination on what they looked like needs to retrogress to the Rice-Burroughs novelette, but our approach to a solution needs to progress. It cannot be the by-gone visual impressions of a more innocent age. The origin of Neanderthals is within us. It's not theories of gradual evolution that are relevant, but it is today's genetic theories— lateral verses lineal speciation.

The difference between the two is vast. Lineal speciation (Gradualism) was the view that species naturally evolve to become more complex species. Lateral speciation basically began with the view of a monk named George Mendel and his Mendelian Bean Bag Theory, that is, that hereditary traits can shift around like beans in a bag. They can come to the surface and also go back in the undertow to come up again in another generation(s). This was greatly different from Charles Darwin's view that variation in a species was microevolution and therefore a minute sign of the process of *trans-speciation* or true macroevolution, which would culminate in one species eventually becoming something entirely new and more complex.

Although Mendel's theory was concurrent with Darwin's, it did not get much attention due to the greater paradigm of the time. Victorian Britain was the age of Uniformitarianism, a suffocating paradigm that favored any kind of gradualism. Ironically, this view had its origins not in biology but in geology. Uniformitarianism was founded on visual impressions based on a dangerous presumption— "the present is the key to the past." The philosophy was the brainchild of James Hutton, a Scottish medical doctor and lawyer who, as a gentleman farmer of the 18th century, formulated the concept of the "Day-Age Theory," better described by his maxim above. In essence, Hutton proposed that the cumulative effects of the Earth's daily weather patterns and current cycles were sufficient, if compounded over time, to create all the geologic features in the world.

In the 19th century, Uniformitarianism finally crystallized in the hands of another barrister turned amateur geologist. In the 1820s Charles Lyell decided to travel around the world, as would become the vogue of "men of erudition," and record his visual impressions of its features. Taking to heart Hutton's maxim, Lyell looked upon the deep sediments of the Earth's crust and assumed that these successive sheets of water-laid rock were laid down slowly over eons of time according to the daily process of nature that we all take for granted each year. This became the basis for his voluminous work *The Principles of Geology*, published in three volumes between 1830 and 1833, with which he became the father of modern geology. For whatever reasons, he could not and would not consider that any great cataclysm had struck the Earth. Because the Bible and other religious texts and ancient histories around the world contradicted this by mentioning such events as a great flood and a great earthquake and the cataclysmic division of continents, the new sophisticates used Lyell's work to delegate these stories to primitive exaggeration and folktales, something that no up-and-coming Bourgeois would wish to embrace.

Nothing now stood in the way of the "new science." Embracing Hutton's maxim, Uniformitarians could only come up with ridiculously long ages to accomplish any geological formation because every minute change noted today was the standard to extrapolate backward to provide a dating method. For example, silting at the mouth of a river is a slow, daily process. Since much of the Earth's crust was also made of such sediment, Uniformitarianism proposed it was created in the same way over eons of time, and therefore it would have taken millions and millions of years to slowly lay down all the sediment covering the continents. Calculating years by the centimeter from bedrock (which was given the name Pre-Cambrian) to the present Cenozoic Era's Quaternary Period stratum, it was deduced that this represented roughly 600 million years of elapsed time.

Despite being a geologic paradigm, the uniformitarian mentality grew to envelope every aspect of Western thought. Progress was viewed as natural, slow and inevitable. It especially overtook biology where all species were subsequently proposed to be the result of the same process, only in this case the slow build-up was that of biologic changes.

Fossils were claimed as proof of such a view. Uniformitarians insisted the lowest strata carried the simplest life forms. Those higher up (in other words those closer to the present-day surface) were fossils of mammals and more complex, advanced life forms. Judging by the slow rate of sedimentation today, the deepest strata must have been about 600,000,000 years old, and the life forms therein obviously of equal age. Therefore the Geologic Age of anything found within the strata could be determined by its depth in the strata. This became Uniformitarianism's Geologic Column. On paper and in theory it was a neat ascending geologic graph that could also double as a biologic calendar of history.

[ 169 ]

Lyell's *Principles of Geology* thus became a crucial support for the theory of gradual, lineal evolution. Uniformitarianism influenced Charles Darwin, who was to become the most famous gradualist, to believe that evolution must also be a very slow, linear process, and the uniformitarian interpretation of the fossil record justified his view that life ever-so-slowly evolved from simple to complex forms. It also gave him the long ages needed if his theory could ever work. In his Introduction to *Origin of Species* he wrote: "I am fully convinced that species are not immutable; but that those belonging to what are called the same genera are lineal descendents of some other and generally extinct species, in the same manner as the acknowledged varieties of any one species are the descendents of that species. Furthermore, I am convinced that natural selection has been the most important, but not exclusive, means of modification."

The concept of evolution was not new, but "natural selection" was, and it sounded as if it was neatly and provocatively explaining the influence of nature on morphological change. In short, it sounded like Darwin had actually discovered the mechanism or an integral part of the process for evolution, especially in light of his comments: "As many more individuals of each species are born than can possibly survive; and as, consequently, there is a frequently recurring struggle for existence, it follows that any being, if it vary however slightly in any manner profitable to itself, under the complex and sometimes varying conditions of life, will have a better chance of surviving, and thus be *naturally selected*. From the strong principles of inheritance, any selected variety will tend to propagate its new and modified form."

It sounded logical, but what exactly did he mean by Natural Selection? Many were not certain. Throughout his *Origin* it was unclear, causing him to pen in later editions of his 4th and pivotal chapter "Natural Selection— Survival of the Fittest" that "Several writers have misapprehended or objected to the term Natural Selection. Some have even imagined that natural selection induces variability, whereas it implies only the preservation of such variations as arise and are beneficial to the being under its conditions of life. . .

Others have objected that the term selection implies conscious choice in the animals which become modified; and it has even been urged that, as plants have no volition, natural selection is not applicable to them! In the literal sense of the word, no doubt, natural selection is a false term; but who ever objected to chemists speaking of the elective affinities of the various elements?— and yet an acid cannot strictly be said to elect the base with which it in preference combines. It has been said that I speak of natural selection as an active power or Deity; but who objects to an author speaking of the attraction of gravity as ruling the movements of the planets? Everyone knows what is meant and is implied by such metaphorical expressions; and they are almost necessary for brevity. So again it is difficult to avoid personifying the word Nature; but I mean by Nature, only the aggregate action and product of many natural laws, and by laws the sequences of events as ascertained by us. With a little familiarity such superficial objections will be forgotten.

Time has proved the objections were not superficial at all. In Darwin admitting that Natural Selection was merely the way to abbreviate the observation that all species showed adaptation to their respective environments, he was admitting he had no mechanism for what had caused them to biologically vary in the first place. Simply put, by what mechanism did those that did vary or acquire some "favorable change" indeed acquire that favorable change to begin with? He had hoped that over long periods of time, coupled with climactic and environmental changes, subtle morphological changes would take place and add up. Although the gravity of it was lost in the public forum, it was undeniably evident that Darwin had *no* mechanism for evolution. Natural Selection was not a creationary mechanism; it was merely a preservational concept.

Over his entire lifetime Darwin took repeated attacks at the foundation of his theory. One particular argument shook him deeply. It was pointed out to him by Fleming Jenkins in the *North British Review* in 1867: "It is impossible that any sort of accidental variation in a single individual, however favorable to life, should be preserved and transmitted by Natural Selection. . .because the advantage, whatever it may be, is utterly out balanced by numerical inferiority . . .such variation would be swamped just as a highly favored white cannot blanch a nation of Negroes."

Molded by his dedication to the pursuit of truth, Darwin knew Jenkins' arguments (by the theories of heredity at the time) were crushing. As a result, in many respects, Darwin had to modify and modify again his thesis until he basically was adopting the theories of Jean Lamarck. Some 40 years before him Lamarck had proposed a gradual naturalistic means of speciation. Darwin had summarized Lamarck's contribution to evolution in his Historical Sketch: "With respect to the means of modification, he attributed something to the direct action of the physical conditions of life, something to the crossing of already existing forms, and much to use and disuse, that is, the effects of habit. To this latter agency he seems to attribute all the beautiful adaptation of nature;— such as the long neck of the giraffe for browsing on the branches of trees."

Lamarck's ideas were essentially that in attempting to reach the higher branches of the trees, giraffes inch-by-inch stretched their necks and these traits passed on to the next generation. Thus, over time, giraffes were born with longer necks. Darwin, however, in 1868, the year after Jenkins' sobering arguments, now proposed a mechanism to explain how Lamarckian concepts of acquired traits could pass on, therefore overcoming the arguments Jenkins put forward about numerical inferiority. It was Pangenesis. Darwin proposed that "gemmules" were given off by each cell of the body and gathered into the sexual organs before fertilization. Thus at procreation it was possible that acquired traits could be passed on to the next generation. Each organ of the body had a number of gemmules based on its size (more cells). As the giraffe's neck grew longer and bigger by habit of stretching it, it acquired more cells and thus more "pangenes" or "gemmules" were sent to the next generation on procreation. Thus the next generation of giraffes was born with longer necks. Over time this amounted to the long neck of modern giraffes.

Pangenesis, however, was essentially destroyed by a scientist named Weissman. He cut the tails off 20 generations of mice and yet each generation was still born with a tail. Obviously, the organs did not carry any such pangenes.

Darwinian Theory was spared the ignominy of the fate of Lamarck's perhaps because Darwin's thesis *Origin of Species* stood separate to all his qualifying works (Pangenesis appeared in his 1868 book *The Variation of Animals and Plants under Domestication*) and was heavy on observations rather than on "how" evolution could happen. As a result his name's synonymy was with evolution in general, not in any detail (beyond Natural Selection) that could be subject to exposure. Lamarckian theory collapsed because it was heavy on speculating "how," and "hows" could eventually be disproven. Ironically, though he had essentially become Lamarckian, Darwin's thesis was spared exposure because his Lamarckian compromises were imbedded into later revisions of *Origins* and thus buried as scattered convolutions amidst his other theorizing. This insulated Darwin at the time, but the result of later study of Darwin's evolving thesis was noted by Loren Finley, who in 1959 wrote *Dar-*

*win's Century:*

> A close examination of the last edition of the *Origin* reveals that in attempting
> on scattered pages to meet the objections being launched against his theories
> the much-labored-upon volume had become contradictory. . .The last repairs to
> the *Origin* reveal. . .how very shaky Darwin's theoretical structure had become.
> His gracious ability to compromise produced some striking inconsistencies. His
> book was already a classic, however, and these deviations for the most part
> passed unnoticed even by his enemies.

The inviolability of "Darwinism" persisted in the public forum largely because of such rabid anti-clerics as Thomas Huxley and Herbert Spencer. Preaching dogmatically the "fact" of Darwinism, Huxley even earned the reputation as "Darwin's Bulldog." Yet it is a fact that Huxley was a bulldog for evolution, not for facts or even for Darwin. Thomas Huxley, even as Julian Huxley wrote of him, wanted to make evolution a "religion without revelation."

For many gradualists this became the way. Evolution was preached rather than proven. But it was only outside of England where science voiced its objections to Darwinian extrapolations the most poignantly. When in 1872 an attempt was made to elevate Darwin to a more established position in the international scientific community (he was essentially a hobbyist in entomology), the prestigious French Institute's Zoological Section voted him down (he only achieved 15 votes out of 48). They declared: "What has closed the doors of the Academy to Mr. Darwin is that the science of those of his books which have made his chief title to fame— the 'Origin of Species,' and still more the 'Descent of Man,' is not science, but a mass of assertions and absolutely gratuitous hypotheses, often evidently fallacious. This kind of publication and these theories are a bad example, which a body that respects itself cannot encourage."[35]

Darwin himself acknowledged that his evidence could be interpreted differently. In his Introduction he even penned with laudable insight: "For I am well aware that scarcely a single point is discussed in this volume on which facts cannot be adduced, often apparently leading to conclusions directly opposite to those at which I have arrived. A fair result can be obtained only by fully stating and balanc-

---

[35] Nevertheless, in 1878 Darwin was put forward as a Corresponding Member in the Botanical Section of the French Institute, and as a sort of compromise did achieve election, prompting even Darwin to quip: "It is rather a good joke that I should be elected in the Botanical section, as the extent of my knowledge is little more than that a daisy is a Compositous plant and a pea is a Leguminous one."

ing the facts and arguments on both sides of each question; and this is here impossible."

Never as dedicated as Darwin, the popular forum saw no such debate and balancing from his followers. Instead Gradualism became the popular nature religion Huxley so wanted. Perpetual evolution was its greatest folk concept. Everything would continue to evolve and get "better." "Missing link" became its greatest artistic fraud. Certain that man evolved from apes, no less than Henry Fairfield Osborn proposed the black African as a genuine missing link. Osborn, a wealthy leading mammalian zoologist and paleontologist, was also head of the American Museum of Natural History in New York. In the 1920s he wrote that an unbiased zoologist from Mars would easily classify people into several genera and distinct species. Indeed, Negroes would be classified as a separate and distinct genus, one not fully evolved to complete human standing. It was a fact of evolution, wrote Osborn, that "The standard of intelligence of the average adult Negro is similar to that of the eleven year old youth of the species *Homo sapiens*"— which for Osborn meant Europeans.

Fortunately, Gradualism's crude visual impressions would soon be challenged. As gradualistic evolutionary theory progressed into the 20th century biologists were painfully aware they still had no unassailable mechanism to explain how a species developed a variation let alone justification for the theory of evolution. But Mendel's Bean Bag Theory was now gaining widespread support and proving itself inspirational. Biology was discovering that the difference in people, no matter how stark the visual contrast might appear, was actually very small. Skin color, for example, was merely caused by the protein melanin. All people were essentially the same color. Those who were darker skinned simply had a greater concentration of melanin.

Genetics became a new science. In the hands of Mendelian theorists, it was probing into heredity and into the origins of where traits came from. Mutations were now becoming understood. This finally gave gradualistic evolution a mechanism: mutation of the gene. The gene can mutate over time, giving rise to new traits. If by chance (literally) this random mutation favored a species in its environment then it was the fittest and would have an edge in its environment, and multiply. It was true evolution based on random mutation and preserved (metaphorically) by Natural Selection.

Mutation/Selection thus became the backbone of Neo-Darwinian Theory, which itself became the dominant and sacrosanct theory in textbooks for the greater part of the 20th century. But Gradualism and the Progress Theory still predominated. Thus it quickly tailored the Mendelian view to support lineal speciation. Yet despite its firm belief, Neo-Darwinian Theory still had no real mechanism for mutation. What caused the mutation? Discovering and eventually unraveling some of the mystery of DNA was immense progress in understanding heredity. But herein was also found the deathblow to Gradualism. The DNA molecule would prove so complex, with its information equivalent to a vast library of books, that even evolutionists couldn't restrain from using creationary language. In his article entitled "Mechanism of Evolution" noted Neo-Darwinist Francisco Ayala paradoxically defined a mutation as an error in the DNA.

This was truly Mutation/Selection and hence Neo-Darwinian Theory's greatest problem: genetics indicated that the chicken came first. Mutation, as a concept, already implied a design preexisted. And yet Darwin's concepts were totally based on gradual evolution being *inherent* in nature.

Darwin had said ". . .I mean by Nature, only the aggregate action and product of many natural laws, and by laws the sequences of events as ascertained by us." But it is a fact that Darwin's gradualism was based on a philosophy: the Progress Theory. Natural laws reveal only one sequence of events— decay. A biologist may look at an organism and consider it mass or matter, but a physicist knows it is energy in another form. No material— whether water, iron, calcium, lithium, carbon, oxygen, etc.— is anything more than neutrons, protons and electrons (except hydrogen which has no neutron). The difference in each substance is therefore the electromagnetic energy created by the varying number of these particles in the *space* of their orbits. In the atomic age we have learned that in a very real sense *all* things are energy.

The reason why "Time's Arrow," as Isaac Asimov put it, points everything toward decay is that *energy is not subject to evolution*. It is subject to the interactions of thermodynamic processes. With this comes *entropy*, a nifty Greek work meaning "in-turning." A fraction of the energy that goes to run any mechanism turns in on it and starts destroying it. This very process is at work in all nature, in our own body systems and in cell division.

The laws of Thermodynamics are some of the most immutable. The 2nd Law of Thermodynamics makes it clear that any system left to itself (closed system) must become more random and disordered. A good example of a closed system is a dead body. While it lived (open system) it took in oxygen, fats, vitamins, water, food; and its mechanism (digestion, respiration, metabolism) converted these into viable energy to run the body. Now that it is dead the mechanism is no longer functioning and the natural sequence of events takes place: the body decomposes and the system (body) becomes more disordered until final decay has erased it to brittle bones.

The mechanism of energy conversion (metabolism) cannot stave off the universality of the 2nd Law of Thermodynamics, for with energy conversion comes the Second Law. Death, in essence, is the end result of a lifetime of our bodies (biologic open systems) trying to stave off the greater universal physical principle of entropy that governs the energy of the matter out of which we are made.

Jeremy Rifken pointed out (Entropy: A New World View). "The entropy law will preside as the ruling paradigm over the next period of history. Albert Einstein said that it is the premier law of all science; Sir Arthur Eddington referred to it as the supreme metaphysical law of the entire universe."

Gradualism is insoluble in such a medium. Rifkin continues:

> We believe that evolution somehow magically creates greater overall value and order on earth. Now that the environment we live in is becoming so dissipated and disordered that it is apparent to the naked eye, we are beginning for the first time to have second thoughts about our views on evolution, progress, and the creation of things of material value . . .Evolution means the creation of larger and larger islands of order at the expense of even greater seas of disorder in the world. There is no single biologist or physicist who can deny this central truth. Yet, who is willing to stand up in the classroom or before a public forum and admit it?

Jeffery S. Wicken noted the contradiction of evolution to the Entropy Law ("The Generation of Complexity in Evolution: a Thermodynamics and Information theoretical discussion," *Journal of Theoretical Biology*, volume 77, page 349.)

> The cosmological arrow generates randomness and disorder, whereas the evolutionary arrow generates complexity. A fully reductionist theory of evolution must demonstrate that the evolutionary arrow can be derived from the cosmological arrow.

This has been the scientific impossibility. A universal theory that posits growing biologic complexity cannot be derived from a universal physical principle which is clearly observed to continually generate disorder.

On top of this is the mystery of life. Life is not chemical. A dead body has the same chemicals in the approximate sequences as a living body; but it is dead. The difference is oxygen. The once treasured and prime energy source has not only ceased to be taken in and utilized, it commences the decomposition of the body. It is oxygen in the atmosphere which causes decay. Without the mechanism of respiration intact and fully functioning, the prime energy source is actually the culprit of total destruction. Energy cannot create the mechanism. Indeed, oxygen is a destroyer not a builder without a mechanism already present to convert it to use.

As Gradualism progressed in the 20th century, it was becoming clearer and clearer there was no science to it. In that halcyon time of "question the establishment," overt admissions of this could merit publication. An eminent British biologist, L. Harrison Matthews, comments in his foreword to the 1971 edition of none other than Darwin's *Origin of Species*:

> The fact of evolution is the backbone of biology, and biology is thus in the peculiar position of being a science founded on an unproved theory— is it then a science or a faith? Belief in the theory of evolution is thus exactly parallel to belief in special creation— both are concepts which their believers know to be true but neither, up to the present, has been capable of proof.

Not only was no proof forthcoming, Gradualism, like many faith systems, incorporated deeply unstable philosophies that contradicted some of its most established tenets. For instance, the standard of "missing link" was predicated on the notion of finding evidence of bipedality in a fossilized primate bone. On the other hand, the gibbon was considered the most primitive of the great apes, and yet it is the only one that is habitually bipedal. The Pygmy Chimp of the Congo spends a great deal of time on its hind legs as well. There is the case of a Celebes Black Ape that learned to walk habitually bipedal in the Hong Kong Zoo in a most un-apelike fashion. What then is the value of finding knee, hip and leg bones in the sediment that indicate by wear or design that the owner could have been bipedal or partially bipedal? There is, in fact, no value in terms of "missing links."

Gradualism also put too much emphasis on the somatic appearance, overlooking that intelligence did not follow its lineal scheme. Although mankind is the most intelligent, apes most certainly are not next in line despite their somatic similarity to us. The next in line is actually an insect.

Ants engage in collective and planned engineering. Some ant nests can go down 30 feet and are marvels of design, literally honey-combed with passages. Ant engineers work under drones and they construct water drains and traps to prevent flooding. Ants are the only other creature on Earth that engage in ranching, cattling aphids for their milk. They are the only other creature that has generals, that has a chain of command and that makes war. They are also the only other creature that makes slave labor of their captives, pressing them to orderly work for the benefit of the nest. Scouts are dispatched in logical patterns to find food. Highways of traffic ship food stuffs from hundreds of yards away to the nest. Compared to the ant, the chimp is a moron.

Bees are themselves an incredibly intelligent insect. Their language is the most translated and interpreted non-human language on Earth. We understand how they communicate distance, direction, and even the quality of a food source. They are the only other creature that actively engages in adjusting their atmosphere artificially. Have you ever wondered what the buzz is from inside a hive on a hot day? They have the air-conditioning on! Bees must keep the larva in the brood area a certain temperature (below 97 and above 90 degrees F). On hot days they gather water and place it strategically throughout the hive. Many bees then line up in advantageous positions and flap their wings. At one end of the hive they are positioned to drag in air; at the other they are propelling it out. It is complete, literal circulation of wind- and water-cooled air through the nest. They did this long before mankind.

Few humans have the wits or ability to spin a spider's web, and yet it knows how by instinct. As geometrically complex as a spider's web is, mankind should be impressed.

Imagine the aforesaid traits in a tribe of apes. They could build complex dwellings, place guards to protect their "villages," have generals to conduct war. They could even take humans captive (also primates) after raiding a farm and press them to work cleaning the village. They would have a king or queen, stock pens for cattle, perhaps even construct a primitive air-conditioning system with a network of punkers. They would milk cows and disseminate the milk amongst the tribe. They could weave nets and capture food. Would they not strike us as people of some kind? But they would not be. Their intelligence and organization would not be in any way more advanced than ants or bees.

The chimp sometimes uses a stick to dig for food (sometimes ants!)

and we declare "tool using"! Yet this is nothing in advance of what one of the Galapagos finches is observed to do with a cactus needle when it pries into bark to get at its meal. The beaver puts them all to shame, using sticks to build an habitable dwelling. Few creatures have the ability to adapt the environment to their benefit. Man is one; ants, bees, even the beaver are others.

Recently on September 6, 2010, psychologists surprised the news. In a briefing before the Supreme Court, Harvard's professor Richard McNally declared of "repressed memory" that it doesn't exist. "It is the most pernicious bit of folklore ever to infect psychology and psychiatry." Griffith University's Grant Devilly said that sufferers of traumatic experiences actually have the opposite problem. "They wish they couldn't think about it." Yet it is a fact that for over 100 years Harvard and every other university has ingrained the whole Freudian concept of repressed memory in its undergrad psychologists. For over 100 years psychologists simply were indoctrinated is an unsubstantiated view and therewith did not notice or listen to their patients when they said something outside the realm of their education.

This is quite analogous to Gradualism's tenets. The world of intelligence never followed some gradualistic pattern, with man at the top and apes next. Yet for over 100 years biologists and zoologists have studied apes, especially the chimp, and marveled at their intelligence when in reality they are back-aching morons who would be hard-pressed to match an ant in industry, a bee in intelligence, and a house cat in cleanliness. Intelligence in an ape was determined by how it could be trained to do simple things, like an automaton. Yet is that intelligence? In human beings the most intelligent are not mindless lackeys.

In 1973, Pierre-Paul Grassé, considered the "Dean of French zoologists," was one of the first to finally condemned Darwinism. Holder of the Chair of Evolution at the Sorbonne University, he declared in his book *L'Evolution du Vivant*: "Through use and abuse of hidden postulates of bold, often ill-founded extrapolations, a pseudoscience has been created. Biochemists and biologists who adhere blindly to the Darwinist theory search for results that will be in agreement with their theories." At best he considered them naïve. At worst that they assume the Darwinian hypothesis is true and therefore make the evidence fit. "This deceit," he continued, "is sometimes unconscious, but not always, since some people, owing to their sectarianism, purposely overlook reality and refuse to acknowledge the inadequacies and the falsity of their beliefs."

Against Neo Darwinism's use of mutation, Grassé noted that mutations can do no more than cause variations within a type. They are not "cumulative" or "complimentary" and they only "modify what preexists." In short, they are "merely hereditary fluctuations around a median position; a swing to the right, a swing to the left, but no final evolutionary effect."

By Grassé's time, biologists had discovered why the DNA was not capable of allowing huge shifts. DNA has a self-protecting mechanism that repairs sequence replication and only rarely does an uncorrected sequence in the chain pass through. This, the true mutation Neo Darwinists had sought was, however, truly an error in a programmed chain of complex chemicals. Remarkably, DNA understood its code and prevented evolution by repairing errors. Mutation was actually something the molecule fought against. This made mutation a fairly rare occurrence. Only once on an average in about $10^7$ replications of a DNA molecule will an uncorrected mutation get by.

Mathematically that proved daunting to orthodox Neo-Darwinian Theory. One mutation isn't going to build a new structure, not even 2 mutations will do that. *And in order to do so* the mutations must be *sequential*. Each mutation must build upon the other logically in the chain of instructions and apply to a specific organ. It would be phenomenal to even consider 3 sequential mutations since that would require $10^7$ x 3 or $10^{21}$ of hypothetical genetic material in which such a mathematic probability could be conceivable. Such a number, one hundred billion trillion is more volume than ocean water on this planet. Four sequential mutations would mathematically require $10^{28}$ amount of material— more matter than the Earth. Thus the possibilities for evolving life by any slow Neo-Darwinian concept of mutation/selection was literally impossible— ludicrous when it was computed that there are a little less than $10^{80}$ atoms in the known universe. Invoking time—i.e. long ages— didn't help. Five billion years, the pet number of Neo-Darwinists, only amounts to $10^{17}$ seconds. By the 1960s it was plain that mathematically orthodox Neo-Darwinian theory was a shambles.

In 1967 the nail was put in its coffin. The Wistar Institute saw a gathering of some of the most prestigious mathematicians and biologist in the world to discuss the problems in the then-dominant theory. Many of the French shared Grassé's yet-unpublished views. The paper from the University of Paris' expert, Marcel Shutzenberger, caused several outbursts of emotions. The chairman, C.H. Wadding-ton, even openly declared "Your argument is simply that life must have come about by special creation!" The record of the minutes shows Shutzenberger and the audience repeatedly responding "No, No!" Nevertheless, the findings of Shutzenberger's *Mathematic Challenges to the Neo Darwinian Interpretation of Evolution* were a deathblow, and even Waddington later declared himself a Post-Neo-Darwinist.

Pray, what is that? To this day it has truly never crystallized. It, too, is drenched and diluted by the popular forum. As you may guess, it is certainly nothing that is supposed to be gradualistic. The most famous Post Neo-Darwinists have been Stephen Jay Gould and Niles Eldridge. Punctuated Equalibria, their theory of saltatory or evolution in big jumps, had several insurmountable problems as well, which Neo-Darwinists were quick to point out. Of these problems, mathematics was definitely a major issue. Just how many dozens upon dozens of mutations would have to come about at one time to permit the sudden saltatory evolution of a given specimen in a species so that it becomes something else? The mathematic impossibilities are even more obvious than Neo-Darwinian Theory since long periods of time are removed. Francisco Ayala raised them, posing the question that if a lizard's egg did phenomenally hatch a bird just where would it find its mate? Just how often does this happen?

Gould, however, had the position at Harvard and even more importantly the dynamic personality to be of command to a media that cares little for the substance or probability of scientific theories. For years Gould poked fun at Neo-Darwinian Theory and then in 1980 in *Paleobiology* "Is a New and General Theory of Evolution Emerging?" declared of orthodox Neo-Darwinism "That theory, as a general proposition is effectively dead, despite its continuance as textbook orthodoxy."

Gould's Punctuated Equilibria, however, did not replace Neo-Darwinism. It is one of many Post-Neo Darwinian theories that vie for recognition. Probably most accurately put, Punctuation has become a favorable concept for speciation. But as with Neo-Darwinian Theory it shares the same problem of having no actual stimulus. And like Neo-Darwinian mutation/selection, it has as its obstacle the orderly function of DNA.

Discovery after discovery has revealed DNA as beyond complex, with many tiers of functions, the last being beyond even the furthest reaches of biology. If biology is viewed as the study of an organism's hardware, then DNA reveals there is actually a software programming instilled in organisms that goes beyond the gene codes. When the chain of nucleotides centered on the bases adenine, thymine, cytosine and guanine that form thousands of letters that are the coded instructions for the cells is found to have a nucleotide out of place the DNA molecule will actually try and fix it. That indicates a higher supervising programming monitoring the gene sequences.

Both in reproduction and daily function DNA is remarkably conservative. DNA's immense self-protection programming is, in fact, a marvel of maintaining equilibrium. On an average, DNA repairs up to 1,000,000 molecular lesions per cell per day. As enormous as this may seem, it is actually a minor percent of the vast 3 billion base pairs of the human genome. It is only when damage remains unfixed and then the new, mutated chain of DNA is replicated that the original code is lost by the daughter cells.

Minor tampering with DNA replication therefore cannot create punctuated surges in speciation; but messing with that 'supervising programming' theoretically could since it controls the repairing. To mess with the actual chain is to mess with gene traits; but to mess with the mechanism that keeps them in check is to remove the mechanism that repairs 1,000,000 lesions per day. Without it dozens of mutations might possibly come about in a chain.

Pursuing this angle perhaps might provide the most potential support for the entire concept of Punctuation. The missing factor is not how but what. What could so interfere with the rhythm and programming of the DNA molecule that it could go haywire for brief periods allowing a radically different species to suddenly emerge? Logically, there seems no other option but a force that can upset the very atomic structure and its electromagnetic rhythm; all vital components, naturally, of the DNA. At the same time the supervising programming is dysfunctional, the reproduction in the chain must also be disturbed, for enough specimens of a species would have to undergo a similar or identical change in order to reproduce a viable new breeding population. Therefore the stimulus to the DNA shakeup would have to be external and widespread, affecting populations and not just an individual. Widespread basically means it would have to affect huge areas of the globe in order to encompass enough specimens.

Unfortunately, Gould was not entirely that big on global cataclysm. Catastrophism was the haunt of creationists, a "sect" that Gould liked to poke fun at. To his chagrin, perhaps, discoveries by his chief opponents might be the only thing to provide his theory with any viability, at least for speciation.

Uniformitarianism did away with the whole concept of global cataclysm purely on philosophical grounds and its visual impressions. A global flood, so embedded in the memories of all ancient peoples, seemed impossible given modern knowledge of all the land masses and the height of the mountains. Where could the water have come from and whence could it have gone? What could have caused it to begin with?

However difficult it may be, it has not proven impossible. By the time that Neo-Darwinian Gradualism was breaking up in biology, Uniformitarianism was also breaking apart. Ironically, it began with the very foundation of Uniformitarianism's entire theory— the strata of the Earth. Sediment is water or wind laid rock. Three quarters of the Earth is covered with it. In some places it is over a mile deep. The fact that entire (and huge) animals were found buried in it does not argue for a slow, thousands of years, build-up of sediment, but a sudden and vast deposit— in other words, a *flood*.

Even the source of a great flood seems revealed in the sedimentary layers. Worldwide geologic corings have revealed flora and fauna in every location, including Antarctica, that thrive in tropical regions, indicating that the Earth had had a steady greenhouse atmosphere at one point. Only three things can cause that: more ozone, carbon dioxide, or water in the atmosphere. Taking into account the current amount of water on Earth embodied in the polar ice fields, which if it melted it would flood the Earth, the cause of such a greenhouse environment had to have been more ozone or carbon dioxide. Otherwise the Earth would have been one massive and perpetual ocean. However, if a greenhouse effect was caused by a water vapor canopy in the Troposphere, then everything works. A collapse of this water vapor canopy would have brought huge amounts of water upon the Earth.

Knowledge of such an event does seem reflected in some of the most ancient writings. In the book of Genesis there is a passing reference to the antediluvian environment. Genesis 2:5-6 reads that "the Lord God had not caused it to rain upon the Earth, but there went up a mist from the Earth, and watered the whole face of the ground." Dr. John Morris, well known "flood geologist" and creationist, proposes that this clearly describes the dew process that would have existed in such an environment. A water-vapor canopy would have created a steady temperature in which there could have been no air-mass circulation and therefore there could be no rain. The temperature drop at night would have brought down some of the water as dew, then with the heat of day it would rise, watering the vegetation in the process. In the passage dealing with abatement of the Flood further corroborating information was found. It is recorded that a great wind came— something noted perhaps because it was the first time a strong wind had ever been experienced by mankind. This would indicate that the atmosphere of the Earth had undergone a great change and now air-mass circulation was possible. With far cooler temperatures, the formation of polar ice and an "ice age" was possible and the waters would assuage.

In any event, enough study of the crust of the Earth has shown that the Geologic Column, the soul of uniformitarianism, actually points not to uniformity but to an horrendous series of cataclysms beginning with flood and ending with massive continental upheaval. More respect should be given to Biblical histories (and others), since these are the same series of events embedded in the earliest histories of mankind as having happened *recently*.

The Geologic Column concept is represented in the illustration opposite. Between the cross-section of two rock strata is an example of the ideal Uniformitarian biologic model of Gradualism— how life supposedly gradually evolved and has become more complex and progressed throughout geologic history. Beginning with the Cambrian Period containing the first forms of recognizable life, such as trilobites, clams and snails, it proposes that each succeeding rock layer contains more complex sea creatures in fossilized form, then terrestrial creatures including mammals and finally man and current species with which we are familiar today. Ideally then, the Cambrian should be followed by the Ordovician Period, which in turn is followed by the others in ascending order: the Silurian, Devonian, Mississippian, Pennsylvanian, the Permian, Triassic, Jurassic, Cretaceous, Tertiary and then Quaternary.

Seldom, however, was such a neat ascending column actually found in nature. The strata at left in the illustration opposite reflect the Kelowna, Kansas, strata. Note that the Triassic sits right on the pre-Cambrian. The strata are dated to these periods based on the fossils found in them. Yet both strata show only depositional action and no erosion. One was laid down right after the other.

The dating of rock via fossils and fossils via the rock was always a strange and dangerously cyclical method. It was based merely on Uniformitarianism's *assumed* gradualism. Wherever a "simple" form of life was found that rock layer was declared one of the earliest, even if that rock layer was on the surface. This is because some fossils are so identified with a period in history (by this concept) that they became Index Fossils, which essentially means they immediately date any rock to the period with which they are identified in the Column. This is regardless if they are found on the surface or

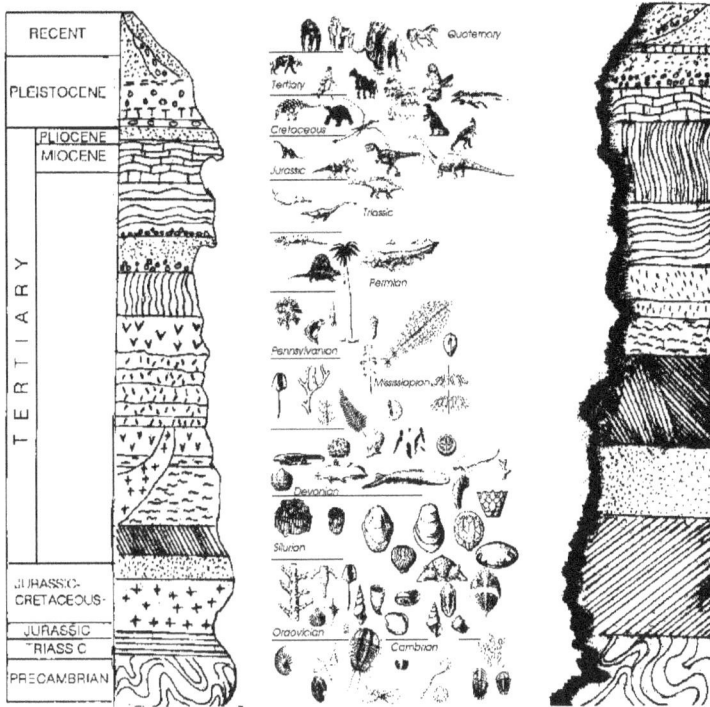

wherever. For example, a primate's bones would date the strata to a recent epoch; a trilobite to the Cambrian Period. But the very gradualistic progression that tells us trilobites were early and primates recent is dependent on the geologic concept of slow worldwide silting being true. If all these creatures were buried in a flood or floods, they all could have been living at the same time and deposited randomly.

Too many examples of randomness exist to support the gradualist interpretation. "Ages" are frequently found misplaced or missing. In some places the Silurian may be missing or the Cretaceous found sitting on top of the Devonian or in one instance the Devonian inverted and coming before the Silurian. This has always been impossible for Uniformitarianism to explain, one reason why in the last 30 years the theory has finally become defunct. In reality, the Geologic Column existed only on paper, but its neat order was never seen in nature.

Flood geologists have come a long way in explaining the order (and apparent disorder) of fossils in the strata. It is to be noted that only sea life appears most commonly on the lowest strata of the "Column," consistent with a flood sweeping ocean water over land,

and not with the sudden explosion of life in the form of what is actually genetically very complex sea life.[36] The process of a huge flood would continue by sweeping over the land larger fish and larger marine mammals and reptiles which could still continue to swim and would not, except in certain cases, be caught in undertows. When they would be, they would be the next creatures deposited in the rising sedimentation. As the waters continued to rise even higher, land animals would continue their escape to higher ground. Those more agile and nimble would gain the highest ground on the hills, while turtles, lizards and large reptilian dinosaurs would not have the speed to do so and would be drowned next and deposited in the rising sedimentation. As the flood continued, those agile mammals on the hills would be the last to be swept away by the tidal currents and therefore the last to be buried in the quickly rising sedimentation.

This is precisely the general layout of creatures in the strata: small marine creatures and crustaceans, larger fish, sea flora, then land flora and small reptiles, then the larger dinosaurs and sea creatures and finally the much more common and agile animals we know of today, aside from apes and man.

The cataclysmic cause of the fossiliferous sediments covering the globe is reinforced by continuing discoveries, many of which were predicted from the Flood Geology Model for interpretation of the sediments. Some of these are, of course, the many instances where entire mountains of fossils have been found. In comment on some of these, Dr. Edwin Colbert, perhaps one of the foremost dinosaur experts of his time, declared concerning a discovery in New Mexico: "As the layer was exposed (the workers had cut a large scallop in the hillside) it revealed a most remarkable dinosaurian graveyard in which there were literally scores of skeletons one on top of another and interlaced with one another. It would appear that some local catastrophe had overtaken these dinosaurs, so that they all died together and were buried together."

---

[36] Dr. Daniel I. Axelrod in his *Science* article "Early Cambrian Marine Fauna" wrote: "One of the major unsolved problems of geology and evolution is the occurrence of diversified multicellular marine invertebrates in Lower Cambrian rocks and their absence in rocks of greater age. These early Cambrian fossils included porifera, coelenterates, brachiopods, mollusca, echinoids, and anthropods. Their high degree of organization clearly indicates that a long period of evolution preceded their appearance in the record. However, when we turn to examine the pre-Cambrian rocks for the forerunners of these Early Cambrian fossils, they are no where to be found." To this phenomenon Dr. T. Neville George added that, granting evolution occurred, "the absence of any record whatsoever of a single member of any of the phyla in the Precambrian rocks remains as inexplicable on orthodox grounds as it was to Darwin."

As worldwide research continued, similar discoveries were being found almost everywhere. Regarding a find in Wyoming, Dr. Colbert noted: "At this spot the fossil hunters found a hillside literally covered with large fragments of dinosaur bones. . .In short, it was a veritable mine of dinosaur bones. . .The concentration of the fossils was remarkable; they were piled in like logs in a jam." Yet again, concerning a discovery in Alberta, Canada, he observed: "Innumerable bones and many fine skeletons of dinosaurs and other associated reptiles have been quarried from these badlands, particularly in the 15-mile stretch of river to the east of Steveville, a stretch that is a veritable dinosaurian graveyard." As far apart as Europe, Russia, Alaska, South America and Mongolia such mountains of fossils have been found. Concerning one found in Belgium, Dr. Colbert noted with impressed tone: "Thus it could be seen that the fossil boneyard was evidently one of gigantic proportions, especially notable because of its vertical extension through more than a hundred feet of rock."

Study of the geo-strata has shown that these were hardly local events. Many geologic formations cover vast areas of entire continents. One such example is the Morris Formation over the mountain west of the United States (with all the dinosaurs fossils in it), and there is also the St. Peter's Sandstone, a glass sand that stretches from Texas to Canada and from the Rockies to the Appalachians.

Many other examples of fossil "graveyards" include the Cumberland Bone Cave of Maryland which contains fossils from a wide variety of species, from bats to mastodons and several species which are inimical— all of which indicates animals fleeing from flood waters or that all their corpses were deposited in the same place by the same tidal current. Dr. Robert Broom estimated that the Karoo Supergroup of Africa contains the fossils of 800 billion vertebrate animals, while the Miocene shales fossil bed of California indicates that one billion fish died within a 4 square-mile zone.

Sudden death is a common feature in the fossil record. In one case, an imprint in fossiliferous rock, all that is left after some buried creatures decayed, revealed the image of a fish half devouring its meal— another small fish— whose tail is still sticking out of its gaping mouth. Fossils have been unearthed of animals caught while giving birth or, as in a herd of Triceratops, discovered by their broken legs and bones to have been overwhelmed by a wall of water while stampeding, no doubt trying to escape. When Trachodon, a dinosaur discovered in sediment formation in Kansas in 1908 (now in the American Museum of Natural History in New York), was unearthed its impression was largely intact. Its skin shows remarkable detail, proving the lack of any significant signs of decomposition or predation by other carnivores and carrion eaters. In August of 1971 an expedition unearthed two fossilized skeletons in the Gobi Desert. The dinosaurs Veliciropator and Protoceratops were still locked in mortal combat, with their backs still arched, indicating that while fighting a sudden load of sediment overwhelmed them and froze them in their death grip.

Another example of how quickly such a flood struck, and how widespread it was, has been found in an unusual place. The huge Colorado Plateau contains several areas where fossilized trees have been found. Like all buried forests around the world these logs showed the same peculiar effects of having been ripped from their place before being re-deposited in the building sedimentation (which trees, of course, do not grow in). Their roots only extend about 3 feet; they have no bark; and, once again, they are sitting buried in sediment— water-laid rock.

But among those in the Colorado Plateau there was an added piece of atomic evidence uncovered to show conclusively that such vast geologic structures were formed quickly. This is the case of "radio haloes" found in the coalified logs. The sediments here are rich in uranium. One of the isotopes in the radioactive decay chain of uranium-to-lead was found to have leached quickly into the logs. This isotope, Polonium 210, has a radioactive decay life of only around 4 years. This decay, as with other isotopes, will shoot an identifiable halo, or concentric shadow or ring rippling out from its core, into whatever substance it is lodged in. In the case of Polonium 210, there is only one concentric ring emitted during decay. The discovery of Polonium 210 halos in the logs showed that the core is surrounded by both an elliptic and a concentric halo.

The conclusion of Dr. Robert Gentry, who has devoted himself to the study, has not been challenged: that the logs were quickly covered by enough weight to compress them and thus the initial halo emitted by the decay; furthermore, after this compression there was still enough time in Polonium 210's radioactive decay life to form another halo, this time one that remained concentric. This discovery indicates that the huge Colorado Plateau was created quickly. The result was to bury part of an antediluvian forest by a load of sediment sufficient to compress the tree logs before the radioactive decay life of Polonium 210 was finished and its ability to create another halo— i.e. under 4 years time.

Field research has even proven that Lyell's great cornerstone of Uniformitarianism— the Niagara Falls gorge— was a blatantly false calculation. Lyell had proposed that the falls receded by 1 foot per year, due to both the pounding action of the water creating the gorge and its swift current eroding the cliff sediments. The math was easy: since the gorge was 35,000 feet long, it had taken 35,000 years for the falls to be at their present position, much longer than the implied age of the Earth according to the Bible. This is what opened many eyes in Victorian Britain to the notion of Uniformitarianism. Despite a geologist of the time named Blackwell telling him in person that the locals had observed the falls recede by over 3 feet per year, making the falls, by Lyell's own uniformitarian criterion, only around 12,000 years old, Lyell stuck to his simple foot per year equation. Modern studies, however, indicate that the Falls actually recede between 4 to 5 feet per year, reducing its age to around 5,000 years.

Uniformitarian estimates for the Grand Canyon were patently inaccurate. Assuming the Colorado River etched the canyon, it was proposed a huge period of time was required. However, then it was discovered that the headwaters of the Colorado River are actually at a lower altitude than the plateau through which the canyon is cut. Since water does not run uphill, it was then proposed that a breach occurred and an inland sea emptied through here. There seems little doubt that the American Southwest was once the bottom of a huge inland sea, and it is through here that it possibly drained out into the Gulf of Baja California. Such a torrent as an inland sea cutting through the canyon would have etched its way through quickly. If the estimates for the Niagara Falls are now genuine, then it indicates that something 5,000 years ago sufficiently jolted the North America continent enough to drop huge sections of it in the Northeast and split apart the Southwest so that an inland sea emptied.

A modern calamity helps to impress upon us how enormous the event must have been that created these geologic features. Mount St. Helens' eruption in May 1980 proved instrumental in finally burying Uniformitarianism. The eruption of the mountain and the attendant earthquakes caused several major geologic events to happen that formed topographical features overnight or in a few days that Uniformitarianism had always declared could only be caused by long ages of slow natural erosion and build up. Vapor winds laid down a 60 foot ridge of sediment overnight. A Gravity Collapse also happened overnight, an area where the ground literally sank to become a flat bottom valley. The lip of the new valley looked weathered as if thousands of years old because the collapse was not even but left many "fingers" extending from the lip to the floor, those familiar grooves and troughs and natural sloughs in the inclining sides. Waves leapt so high from Spirit Lake against the part of the sloping mountain side that sheer cliffs were formed appearing very much like those all over the world that were assumed to be millions of years old. A mud slide passed through a forest and continued on, creating in its wake a 1/5 scale of the Grand Canyon, diverting a creek which then ran down its center.

Years later an entire forest of standing trees was found at the bottom of Spirit Lake. When the volcano exploded in 1980 the force of the blast had ripped about a million trees out of the ground and flung thousands of them into Spirit Lake. Over time this vast log mat had floated and shifted back and forth across the lake with the wind, the bark rubbing off the trees in the process. Eventually the logs became waterlogged and sank, usually root-end first (heaviest end). There they stood on the bottom like a ghostly dead sunken forest. When on team of geologists investigated the lake (headed by Dr. Steve Austin) the base of the tree trunks were securely embedded in the sediment that had rained down and was still building the bottom level.

Entire forests have been found buried around the world, with tree trunks found at various levels of superimposition ascending through thousands of feet of sedimentary rock. Petrified Ridge in the Dakotas is one such famous location. Uniformitarians declared this to be representative of hundreds of millions of years of forestation slowly fossilized over eons by changing ground levels. The problem with Uniformitarianism's rather unusual deduction was that trees don't grow in sediment, they don't moreover grow without their bark and with only 3-foot roots. All this was noted in the fossilized trees of Petrified Ridge. One tree was even found upside down in the sediment.

Careful examination of the "sunken forest" at Spirit Lake proved that here too like Petrified Ridge the trees were denuded of their bark, which had rubbed off as the logs chafed each other while floating back and forth over the surface. The trunks' roots also extended only by about 3 feet since, of course, they had been violently ripped from the ground by the force of the blast. In every detail the trees matched those of Petrified Ridge and other places where hundreds of feet of sediment have been found to be laced with successive layers of tree trunks. From the example of Spirit Lake it is more logical to assume that these vast "petrified forests" were embedded at various levels of the rising sediments based on when the logs became waterlogged and sank, root end first usually. Based on the phenomenon of radio halves in coalified logs deep in the Colorado Plateau we know that this took well under 4 years time.

Spirit Lake's bottom was also strewn with tree bark, the residue of the logs chafing overhead, the process forming a large coal bed. It is a fact that huge pieces of coalified bark have been found perfectly intact at coal sites around the globe, indicating that many of these coal beds have actually been formed by the residue of huge log mats overhead and not by slow uniformitarian theory of peat deposits.[37]

Of course, both points above— forests of fossilized tree trunks and huge beds of coal (which is the fossilized wooded part of vegetation)— indicate log mats floating over sections of entire continents or even over large parts of the entire globe. Naturally, the latter implies only a few survivors, such as in the story of Noah's Ark, which, by the way, is almost universal in some form or another amongst almost all ancient peoples on Earth. The universality of the same story would make sense, equally naturally, since all peoples would be descended of the few survivors and would have carried with them in their migration oral histories of the event.

While creationists effectively destroyed the old Uniformitarianism, the belief is still widely held (and unfortunately taught) and popularly embraced due to antiquated textbooks. The eminent Dr. David Raup, of the University of Chicago, comments:

> . . .contemporary geologists and paleontologists now generally accept catastrophe as a 'way of life' although they may avoid the word catastrophe. . .The periods of relative quiet contribute only a small part of the record. The days are almost gone when a geologist looks at such a sequence, measures its thickness, estimates the total amount of elapsed time, and then divides one by the other to compute the rate of deposition in centimeters per thousand years. The nineteenth century idea of uniformitarianism and gradualism still exist in popular treatments of geology, in some museum exhibits, and in lower level textbooks. . . one can hardly blame the creationists for having the idea that the conventional wisdom in geology is still a noncatastrophic one.

Despite catastrophism firmly reestablished, the popular forum remains a ghastly quagmire of gradualist folk concepts. There have been very little inroads made, even philosophically, into the public forum about the potential effects of cataclysm on DNA and biologic life. Old Hollywood movies more than anything still inspire people to believe in the absurdity of perpetual and lineal evolution, where biologic life is like hand soap. Once the new improved Palmolive

---

[37] Dr. Steve Austin's PhD dissertation was entitled Floating Log Mat Model.

comes out, the old is retired from the shelves. One species neatly melds into a new improved version and lo and behold the old is gone. It is this mindset which makes people marvel that a Neanderthal could still be around. Many still think that we evolved in slow steps from hairy cavemen. Thus if we are here how can they possibly still be alive?

Such a view does a great disservice even to Gradualism's actual thesis. It had proposed that specimens of a species underwent a minute variation and then reproduced more of their own kind. Over time, and compounded with more changes, this amounted to a new species. This never required that the original specie cease to reproduce itself or that it must become extinct. Extinction is simply not a biologic probability if the species is adapted. The problem with Gradualism was its infernally slow concept of speciation, one that was inspired merely by Uniformitarian philosophy. Punctuation, on the other hand, seems far more probable. In Punctuation, a burst of disruption is followed by a calm in which DNA takes up its daily methodical routine again of protecting and reproducing its new chain of instructions. Gradualism's other great problem was that there were no missing links and yet logically by its premises there should be many alive at one time for *every* species on Earth. Gould's great service to the establishment was to impose upon biologists that Paleontology (his specialty) had no missing links. There had, in fact, been no missing links. There were none today and none in all of history.

Mathematically and biologically we now know why. For although the gene pool has the ability to *laterally* vary around a median pole, a huge gulf separates genera. "Minor" is actually relative if not actually infinitesimal when it is viewed in the context of 3 billion base pairs in the DNA. For example, genetically the chimp is only 1.6 percent different than mankind. That seems small, but it is actually a whopping 48,000,000 or so base pairs difference. Consider the difference between mankind and a chimp and imagine what must be contained in that seemingly minute 1.6 percent difference. With the known mutation rate and DNA's propensity to protect itself that gulf can never be crossed. The genetic difference between a "modern man" and a Neanderthal, however, is practically negligible by comparison, each within the reach of the same median pole.

Comparative studies have responded appropriately and cut out the heart of Gradualism. Anthropology has gone back and reclassified Neanderthal Man as *Homo sapiens Neanderthalensis* rather than simply *Homo Neanderthalensis*. He's us. Just another extreme variation like many types of people today. Even poor *Homo erectus* himself has now bit the dust. Based on a minimal amount of bones and a skull cap, remains were declared representative of the first "walking man" species. More finds from Africa and those in Australia, however, show skeletal features indistinguishable from modern humans. Professor William Laughlin, of the University of Connecticut, finally compared *Homo erectus* skull finds with those of living Inuits and of peoples of the Aleut Islands, discovering both of them are remarkably identical. "When we consider the vast differences that exist between remote groups such as Eskimos and Bushmen, who are known to belong to the single species of *Homo sapiens*, it seems justifiable to conclude that *Sinanthropus* [*Homo erectus*] belongs within this same diverse species."

Far more influential on the matter was the Senckenberg Conference in 2000 AD. A "flaming debate" started by University of Michigan's Milford Wolpoff and University of Canberra's Alan Thorne and their colleagues broke out to finally broach the point of classifying *Homo erectus*. "They argued forcefully that *Homo erectus* had no validity as a species and should be eliminated altogether. All members of the genus Homo, from about 2 million years ago to the present, were one highly variable, widely spread species, *Homo sapiens*, with no natural breaks or subdivisions. The subject of the conference, *Homo erectus*, didn't exist."

With this, *Gigantopithecus* as a giant *Homo erectus* and missing link also sails down the plumbing. A giant ape's jaw and scattered fossilized teeth become just that. With a 2 million year old date for modern man, it is clear that both *Gigantopithecus* and ourselves lived alongside each other. There is no way to link human bones to scattered teeth in a curio shop a continent away and pronounce a hybrid evolutionary species. Finds such as *Meganthropus* were thought to be a giant *Homo erectus*. What do these bones now represent?— merely giant human beings?

Reliance on some of the old uniformitarian dating methods led to the date of 2 million years old, as mentioned above, based on the level of sediments of some of the finds already discussed. But even if this remarkable tenure for modern man becomes universally established, it would only raise the questions about recent global cataclysm. For if mankind was on this Earth, the same as he is today, for 2 million years now, then why did civilization suddenly begin only in the last 5,000 years? In itself a 2 million year old date for *Homo sapiens* is amazing, for this date asserts that modern man has been of the equivalent intelligence as he is today for 2 million years, the age of when gradualists had us picturing ape men roaming the globe.

Population statistics add unexpected support for catastrophism. They indicate that humankind was reduced to practically zero around 3300 BC., the time our ancient histories tell us of global upheaval and few survivors. From the survivors of this era all human beings and animals must come. Yet we, as many other species, are a species of diverse and many times radical variations. Therefore from this time period forward all humankind and *all* its diverse variations must originate.

This theory— essentially Punctuated Speciation— is very fascinating given modern knowledge of DNA. Even though the theory is in its infancy, it is worth at least to theoretically ponder these events to see what could have shaken up the DNA and caused any number of punctuated surges and mutations. It may explain what our own ancient forefathers declared about a time before rain and mountains, and a time when strange hairy men were born.

# CHAPTER 9

## Memories of the Future
## Mysteries of the Past

In the last 5,000 years mankind has advanced from brick and stone builders in Sumeria to detonators of the atomic bomb. We have in this short time become master engineers. We have placed a man on the Moon, and now our eyes no longer wistfully contemplate Mars. We make solid plans to send a manned flight to the red planet. In 5,000 years we have stepped from Sumeria into the solar system. Despite formula pundits, mankind has proven himself a very ingenuitive species. Knowledge has built upon knowledge, and each generation is capable of processing the information and using it to take it a step further.

In the face of current speculation that modern man goes back 2 million years, or even if we go back just one hundred thousand years, this is an enormous conundrum. How is it that civilization uniformly begins a mere 5,000 years ago? History, by modern reckoning, dates only to 3100 BC. A mere 5,000 years ago our history begins with the complex written language of Sumerian, recording the religious incantations and legal concepts of the city-states of the Tigris and Euphrates. Around this same time the civilization of the Indus emerges, eventually building cities like Mohenjo-Daro, built on a grid and predesigned to accommodate 40 to 50,000 people.

Egypt also begins around this same time, striding into history building great stone buildings and eventually pyramids. Each civilization grouped around a major river, each shared similarities with the other, each had a similar science and pictographic script, though each spoke a very different language. This date for the division of history/prehistory does not just apply to the "Old World;" the date of 3114 BC is also the date at which Mayan calendars will eventually stop when we calculate backward.

Each civilization shares more than a pictographic script and the coincidence of timing in their emergence. Each recalls a previous era of civilization before them, and each recalls that it was destroyed in massive upheaval. The Maya declare their calendar dates the beginning of a new era, the present world in which we live. According to the Maya as well, the last world was destroyed by massive upheaval.

Not only do these distant civilizations share these histories in common, they also have the same concept for this prehistoric civilization. Each maintains that mankind was one in speech and appearance and that the great destruction of this civilization suddenly divided mankind's language and scattered man upon the Earth. Biblically, the civilization was called Babel, a Hebrew play on words meaning confusion. The Hindi recall *Rutas*, the great civilization that existed before the Himalayas were born and which was destroyed by the upsurge of the great mountain range. Even the remote Cyapo Indians of the Amazon surprised Spanish explorers with their histories of the great corn tree being chopped down and all mankind scattered, a recollection, though in impoverished terms, of Genesis and the Tower of Babel.

Population statistics tell us something very real inspired these blurred histories. It is estimated that 2,000 years ago the Roman Empire had about 100,000,000 citizens. Taking into account all the other peoples of the world, an estimate of 300,000,000 can be made for the population of the Earth. Despite the collapse of the Roman Empire, medieval plagues and poor sanitation, wars, pestilence and famine, mankind is around 7 billion souls today. Yet if mankind were on the Earth for as long as even 40 to 50,000 years, a considerably less span than is now in theoretical vogue, then why aren't there more of the human species— billions and billions more? . . . unless many billions were wiped out long ago in some cataclysm.

The science of population doubling provides an intriguing corollary. For the last 700 years science has been able to determine with a fairly good degree of accuracy how populations have doubled and over how many years this takes place. The first reliable estimate of world population was in 1650 when it was estimated there were 600 million people on Earth. The reason why it took so long to double from the time of the Roman Empire is blamed on the poor conditions of the Middle Ages, Bubonic Plague, Chinese dynastic wars, and the ravages of the mighty and terrible Khan, and from numerous smaller versions throughout succeeding history of "his awfulness." Because of this it took 950 years for the population to double. By 1850, however, the population had doubled to 1.2 billion. Only 100 years later in 1950 it doubled to 2.4 billion. The next doubling was predicted to take only 38 years (1988) in which the world would have 5 billion or more; and indeed the Earth today has far more than that already. Extrapolating these numbers backwards, they stop at around 3300 BC with the estimate of 0. What happened around this general period of time that caused the population to decline to only a few people?

The earliest histories of our ancient forbearers tell us that two great natural upheavals occurred. Evidence for the first— a great flood— has already been discussed in the previous chapter. The second cataclysm happened generations (perhaps more) later after the Earth was repopulated. This was remembered as a terrible earthquake which divided the earth. This latter catastrophe, however, is not so easily reconstructed as is the story of The Great Flood. Even biblically it can only be re-assembled piecemeal. In one reference, the cataclysm happened during the days of the patriarch Eber. In the genealogical recital of Genesis 10:25 it is found written: "And to Eber were born two sons, the name of the one, Peleg, because in his days the earth was divided. . ."

Part of the reason why it is difficult to recognize the intent of some biblical references is the interpretation that has been placed upon them by the traditional clergy. Despite the word "peleg" meaning earthquake, the standard English translation refers to the division of tribes and peoples. This is unfortunate, for it gives the reader a very false impression. The word for "earth" in both Hebrew and Greek (Septuagint) means "ground" and not peoples or tribes which is *goyim, ethnos* or *phyla*.

Another Genesis passage is also indirectly linked to this event, but it too has suffered under translation. The genealogical recital of patriarchs places the era of Peleg as overlapping that of Nimrod and that unknown one-language civilization that was destroyed by God or the gods, who came down and tumbled it. In the statement of "peleg," of course, there is nothing global implied. But the story of

Babel, however thinly at first, implies a global event in the original Hebrew. In Genesis 11:1 the period preceding Babel is introduced and unfortunately translated: "And the whole earth was of one language, and of one speech." This translation has given us a very redundant statement. However, in the original Hebrew the word used for earth is indeed "earth"— *ground.* In the original Hebrew (שפה) does not just mean "lip" but "shore." Scripture's earliest translations into Greek carry over the same literal meaning of ground and shore (Χειλος). In Hebrew "earth" and "lip" are in the singular "and of one speech" is actually plural נדרים אחדים which literally rules out the traditional translation. A correct translation would be: "And the whole earth was of one shore, and all were of one speech." Thus this is perhaps the most overt historic statement that Pangaea— megacontinent— was broken apart during the days of Peleg, which curiously coincided with the era of Nimrod and Babel.

Twofold results followed: sudden language confusion and the scattering of all mankind over the earth.

Putting the two events together is the first step to uncovering the initial chain of events. To the ancients, God or the gods coming down to visit wrath upon them meant something came out of heaven. This suggests an impact— either a meteor, asteroid or comets or shower thereof hitting the Earth. The result of a massive impact or succession of impacts would be a global earthquake.

Questions naturally arise as to how impacts, no matter how severe, and an earthquake, no matter how jolting, could cause language division. The answer is long in coming and can only be tenuously deduced from the events that would have chronologically unfolded after a massive impact. Nevertheless, it is impressive how one effect leads to another and in the end how they corroborate ancient accounts.

To begin with, the first clue is the greatest geophysical effect we have: a planet inclined on its axis. In fact, it is a very clear domino that tells us a devastating chain reaction occurred. The toll spin cannot change. The mantel moves in one direction the crust must move in the opposite. A shift in the axis by 23 and one half degrees is going to shift the mantel and this is going to rip apart the crust. The result is that a shattered Pangaea breaks violently apart. The movement is not a slow, uniformitarian crawl. It is an exponential event, one that starts quickly and slows to a stop. The result is cataclysm, global and horrific.

What almost seems an eye-witness recollection is found in the legend of the Teutonic tribes of Norway. "The whole Earth trembled.

The Yggdrasil tree was shaken from its root to its topmost branches. Mountains crumbled and split from top to bottom. All were driven from their hearth and the human race was swept from the surface of the Earth. The Earth itself was losing shape. The stars were coming adrift from the sky and falling into the gaping void. . . Flames spurted from fissures in the rocks. . .everywhere there was the hissing of steam. And now all the rivers, all the seas arose and overflowed. From every side waves lashed against waves. . .the Earth sank beneath the sea. . .mountains arose anew."

Each statement above could have been inspired by local catastrophes except for that one comment about the stars drifting in the sky and falling behind the horizon. This is exactly how the heavens would appear to those on Earth as the planet heeled over on its axis.

The legend of the Teutonic tribes can be considered an eye-witness description from the vantage point of peoples near the ancient coastline. But the after-effects of continental upheaval from the vantage of those peoples living in central plains would be much different. For instance, it would be daytime for them. They would have no recollection of stars coming adrift. Rather, they would experience an horrendous earthquake, land breaking apart, cities tumbling and mountains thrusting violently upward as the continents buckled at areas of resistance. The stories most familiar to us, originating from Mesopotamia or India, indeed recall just such events.

The Inca recall similar events: the Andes Mountain range violently uplifting and, like the story of Babel or Rutas, they also regaled Spaniards with stories of a fallen city. Harold Wilkins in his *Mysteries of Ancient South America* notes that there were also numerous aboriginal stories which the Spanish and Portuguese explorers recorded concerning ancient recollections of 4 "massive impacts" which hit the Earth and thundered from the distance. These primeval accounts also record afterward that the sky turned various shades of yellow, blue and red. And, most interestingly, in the east, that is, where the Atlantic is now, a great blackness rose *up* from the Earth to the sky.

All of the disaster stories can be variously interpreted, of course, but assuming they were separate events was perhaps the most erroneous interpretation. Few have considered that all these accounts speak of the same single civilization and global cataclysm, and that each account of destruction reflects only the survivors' personal vantage point.

If we are to assemble them logically, the chronology would have to begin with the impacts. This would imply meteor or asteroid collisions with Pangaea where the Atlantic now it. Those peoples closer to the epicenter would naturally remember the impacts and the smoke billowing up where the earth split open while other, more distant tribes from the epicenter just remembered the horrendous earthquake. As Pangaea broke apart and the continents moved, far away on the plains of Shinar (Babel) cities would have fallen. In India cities would have been destroyed by the violent uplifting of mountains as the continental strata buckled in response to tectonic stresses.

It is interesting to note that Plato claimed that Greek knowledge of "Atlantis," a civilization which broke apart and fell into the Atlantic, came from the Egyptians. It is also interesting to note that the name of the Atlantic probably predates any Greek story of Atlas or Atlantis. In fact, in Sanskrit, considered one of the earliest languages, the word for great destruction is *atyantica*. It is further interesting to note that Ignatius Donnelly discovered remnants of Sanskrit in South American languages regarding the early survivors of cataclysm. Does the name of the Atlantic actually come from the earliest languages and does its name actually recall the epicenter of a global cataclysm? Was the myth of an island just a corruption of memories of Pangaea and a great single civilization? Was its destruction— *atyantica*— remembered as its actual name?

Cartographers had long puzzled over the appearance of the continents on opposite sides of the great ocean. In the 16th century, when mapping of the Americas became more extensive, Sir Francis Bacon was the first to note the curious observation that the Americas seemed to fit roughly into the western coast of Africa. This seemed so clear that in 1858 a French scientist named Antonio Snider-Pelligrini actually proposed that the Atlantic had been created by powerful forces that tore apart one great continent.

Proof that the Earth spit open here and belched forth brimstone and smoke was discovered in 1876 when undersea mapping brought to light the Mid-Atlantic Ridge. This ridge is the most curious geologic formation on this planet. It is a massive open wound in the Earth's crust. It has a central rift valley about 2 to 3 miles deep and several miles wide. It extends for about 10,000 miles, and in some places the ridge has uplifted so far from the bottom of the ocean that its peaks form islands as distant from each other as the Azores in the North Atlantic to Tristan da Cunha in the South Atlantic.

The Mid-Atlantic Ridge is not just an unstable underwater mountain range; it has vast inclining slopes. There is no question that the distance between the American and European/African continents is in

direct proportion to the extent of these inclining slopes. Where the
ridge and its slopes are narrow, as between Sierra Leone, Africa, and
Pernambuco, South America, the distance between the continents is
the least. Where it is vast and extends for hundreds of miles below
the ocean, such as between Europe and North America, the distance
between the continents is the greatest.

In essence, mapping has proven that hundreds of thousands of
square miles of the Earth's crust violently thrust upward, spewing
out the contents of the mantel; that which didn't break forth pushed
up and buckled the very skin of this globe. As a result the Mid-
Atlantic Ridge remains the most unstable part of this planet. Thousands of earthquakes strike per year along its open rift valley, and
along its vast inclining slopes there are massive magna flows. On the
bottom of the Atlantic's sea floor there are also gigantic lava beds
that cover areas larger than some countries.

Study of sedimentary deposits has confirmed that some geologic
formations in Europe continue on into Canada and America and the
same can be said for those in Africa and South America. The formations are now, however, divided by the North and South Atlantic
oceans. If the continents could be replaced, like pieces in a puzzle,
the geologic strata would once again connect.

Everything, in fact, seen in the formations of the globe, in oroge-
nie (mountain building), in canyons and gorges and vast beds of ig-
neous rocks, all scarred into the sediments, indicates that the fractur-
ing movement of Pangaea took place after a massive flood which
first laid down all the sediments that were later fractured. This same
chronology is reflected with uncanny accuracy in ancient histories of
our own forefathers.

Furthermore, placing the continental land masses back together at
the Mid-Atlantic Ridge, to theoretically reform Pangaea, and the areas that are now the Caribbean islands would be along the fracture
zone between the two major continents of Africa and the Americas,
near the equator where Pangaea would have undergone the first
crushing impact.

Geologist Walter Alvarez was the first one to popularize the idea
that a huge iron asteroid impacted with the Earth, and even identified a sub-sediment "crater" near Chicxulub, Yucatan (initially
measured at 300 kilometers diameter), that he believed to be the impact site. His theory, though at first ridiculed, later became far more
acceptable, and the fallout from such an asteroid was even blamed
for having wiped out the dinosaurs by having caused an ice age
which would have killed off much of their dietary foods as a result.

In developing this theory it was proven that areas of what are now the Caribbean had been the epicenter for catastrophic impacts that spread forest fires around the globe. This evidence includes Stishovites (shocked grains of quartz) which are most quantitative on Earth in the Caribbean. They have so far only been found at meteorite impact sites and nuclear explosion test beds, where the massive explosions sent shock waves passing through the rock at such force they shock-heat and compress mineral grains of normal quartz.

Intermingled with such massive impact evidence there is the evidence that suggests titanic movement of the American continent in the wake of this impact. Tempestites, or tsunami deposits, have been identified in Mexico, the southeastern United States and in the Caribbean on a colossal average. One was comprised of a 150-300 foot band and of a mixture of both big and small rocks, indicating a violent and tremendous tidal wave ripped through what is now the Caribbean and Gulf of Mexico in the wake of something definitely Earth shaking.

Enormous deposits such as these apparently happened even before the "dust settled." On top of this tsunami deposit a layer of iridium rich silt had settled, and, apparently, has settled over almost the entire globe. Mixed with this metal is carbon ash and soot indicating massive forest fires were burning in response, one large enough to send a soot cloud around the world sufficient to drop a layer of several centimeters in thickness.

Iridium is a trace metal which seeks the presence of iron and is therefore an extremely rare terrestrial metal but one that becomes more common in the mantel and presumably the core where the greatest parts of iron are thought to exist. Its presence today is located in some "hot volcanoes" as those in Hawaii, where it comes up from the mantel in the lava flows. But it is also associated with identified iron meteorite impact sites. This worldwide layer of soot containing the iridium usually measures 2 to 3 centimeters, but in the Caribbean and Haiti it is seen to be as much as 17 centimeters thick with the iridium content being much higher, on an average of 10 to 50 times more.

Concentrations like this would indicate that this was the primary epicenter of the impacts and of the most intense forest fires that followed. This iridium layer contains several "peaks"— i.e. greater periods of concentration— around the globe. A possible explanation for this is one peak caused by the original impact of an iron meteorite and then successive volcanic eruptions caused more peaks as the Earth was jolted and great eruptions sent more iridium into the atmosphere at varying intervals based on the duration and timing of the flows.

Although the theory of an asteroid impact gained wide acceptance, the ridiculously outdated methods of Uniformitarianism were used to date the event back 65 million years. This is particularly ironic considering the impact of such a monstrosity would have shaken continents, caused tidal waves over land and caused any number of areas of sea and land to shift and change places. All of these are events that would build and erode geologic structures quickly, structures that the very philosophy of Uniformitarianism would inaccurately judge to only be capable of forming over vast periods of slow present day weather patterns. [38]

It is more logical to presume that this impact (though not necessarily centered at Chicxulub) happened recently; that it caused the initial breakup of Pangaea and then as the Earth heeled on its axis the continents ripped apart and the Earth split open and belched out clouded death from the exploding Mid-Atlantic Ridge. It is interesting to note that even uniformitarian methodology itself cannot logically assert anything but a very recent break up of Pangaea. Studies of the Earth's atmosphere have proven that carbon 14 is not in equilibrium. Yet the decay rate computes that it should have reached equilibrium in only 30,000 years. Therefore the only conclusion that is tenable is that something happened within this same time frame that shook up the atmosphere. The same can be said for the magnetic field of the Earth. Its decline is the oldest studied geophysical phenomenon. For the last 150 years it has declined by 6%. Extrapolated backward linearly, as is the wont of Uniformitarians, and the Earth would have a magnetic field the strength of a magnetic star in only 7,000 years, which this planet never could have supported. Therefore something must have jolted this planet to the very core where the field is generated in this time frame and started the decline of the field. Even if adhering to Uniformitarianism's precepts of slow silting of the sediments, it is plain to see that the fracture zone between the continents goes up equally between all layers of the Geologic Column, meaning even by uniformitarian standards

---

[38] In his dialogue on Atlantis, Plato records the Egyptian priest speaking of just such a chain of events, which he denigrated the Greeks for preserving as the myth of Phaethon, who "having yoked the steeds of his father's chariot, because he was not able to drive them in the path of his father, burnt up all that was upon the Earth, and was himself destroyed by a thunderbolt. Now this has the form of a myth, but really signifies a declination of heavenly bodies around the earth and a great conflagration on the earth."

that the continents split apart in the last 10,000 years.

Vestiges of a cataclysm that literally shook the world remain as a testimony on every continent. Parts of the Sahara Desert appear to have very recently been an inland sea which for unexplained reasons emptied. Vast lava beds are pierced through the continents in the northwest United States (The Columbian Plateau) covering 200,000 squares miles and of several thousand feet thick; in India the Deccan Traps is a lava flow the size of France. Ayres Rock in Australia, one the largest boulders in the world, is 1,100 feet high with a 5 mile circumference, and seems to have been literally kicked and rolled to its present place in a desert from a mountain 20 miles away.

Most of the greatest mountain ranges of the Europe/Asia and African continents are on their eastern coasts and toward the middle, like the Urals; while it is the opposite for the Americas, where most are on the western side and toward the middle, like the Rockies. The same rings true of the earthquake faults: on Eurasia and Africa they are on the eastern coasts and in the Americas they are on the western, areas that would have born the brunt of the tectonic movement. The continents look very much like a massive carpet that was torn and pushed apart; and as a carpet will wrinkle when it meets with resistance, in like manner the continental strata buckled and mountains uplifted. The same can be said for the bottom of the Pacific Ocean. Off the Americas it looks bunched up like it was being compressed before the American continent and likewise it seems so before China and the Asian subcontinent.

The buckling effect of the Earth's continental surface plainly draws our attention to the Atlantic Ocean, the very place where ancient Amerindians recall giant plumes of smoke belching upward from the wounded Earth and where modern geologic studies have confirmed evidence for massive impact(s).

Cataclysm on this level is rendered invisible by the enormity of its scope, for its evidence *is* the very appearance of the Earth today. Every feature bears some connection to it. But however long it has taken science to uncover the physical features, today we as a reflective culture are only beginning to ask the questions of what could have been the biological effects. Organisms are not stone. They do not solidify and take a form that then lasts for thousands of years and can be back-worked to its origins.

The only clue we have is an oral one. It is said that from this event mankind divided throughout the earth into ethnicities. This was not the immediate effect, of course, as it became evident only with propagation. The immediate effect was a sudden confusion of languages, something that has struck people as folklore. . .until biologic discoveries have uncovered a curious coincidence.

Biology shows us that despite the enormous visual differences between peoples the human race is remarkably closely related. Yet at the same time many human languages show no affinity or contact with another. This is impossible to explain if all languages evolved by a slow, linear progression and then varied along with mankind's dispersal on the globe. Straight to the point, it should be fairly easy to trace all languages back to a mother language, for variant dialects and eventually languages are in themselves a verbal speciation of the same original language. How can a species—*Homo sapiens*— be so closely related and yet have such diverse and unconnected languages?

It is hard to imagine how languages and ethnic appearances slowly came about *a la* gradualism from the same original humans considering that within each ethnicity thousands of years has only produced minor and slow genetic changes in appearances. Ancient Roman busts show the extent that similar features still exist in Italians and many Europeans as they did thousands of years ago. These ethnically distinctive traits, such as nose shape, eyelid shape, body type, whether hair is curly or straight, height, the dreaded fat gene, are all minor genetic variables. But the difference in all five major classes of humanity testifies to radical change in the sequence of nucleotides.

Given that DNA protects itself, we come back to the idea of something radioactive or electromagnetic must be involved in order to have disturbed its routine functions. But the only link we have with cataclysm to this possibility is that odd claim there was a moment of sudden confusion of languages.

There is no denying that the information age has shown us this is an uncanny and prescient statement. The physical of living organisms is merely the hardware that holds a sophisticated world of software programming in which commands, operation and maintenance are carried out by complex electronic signals and codes. Computer technology is the closest thing to helping us visualize our electronic selves. Although its hardware is the only thing we see, this is but the host for the complex electronic signals that integrate and operate every feature it provides. The processor (cerebellum) may assist in the speed with which functions take place. The hard drive (brain) may hold the elusive software codes, the wires (nerves) may transmit them, and the boards (organs) use and direct them, but the electronic codes are controlling the function.

"Software" has really become the only concept that fits and can explain what on the face of it seems inexplicable. Programming, as in any device, has two purposes: to instruct the individual components and integrate them with all other features in the machine. This is exactly what is seen in every living organism.

Even more remarkable are the examples of automated intelligence in the body systems that, though independent of the human will, they nevertheless appear as if they are independently thinking. In each case it involves the integration of sometimes billions of functions. For instance, each cell of the retina does about 10 billion calculations per second in order for sight to occur. Yet as we know the eye is merely a lens that allows enough light in for the brain to capture an image. It is here at the back of our brain that we see, not with our eyes. Like any lens the eyes allow an upside down image into the brain. The brain must invert it. Not only this, but due to the way in which the optic nerve links with the back of the eye, it is impossible to see any image that is about 12 degrees off center from the pupil. There should be therefore a black void in our sight 12 degrees from the center. We are not aware of this only because the brain fills in this area with the surrounding detail so that it is unnoticeable. It is doing this every nanosecond we are shifting our glance. As soon as we have opened our eyes at a given location the brain has read the surrounding details and will constantly keep the area filled in with the appropriate and accurate detail as our eyes shift. If looking at a brick wall, for instance, the brain actually fills in the spot of blind vision brick-by-brick with the actual bricks because it has immediately seen and computed the whole wall. This is literally a mind-boggling example of RAM— random access memory— that defies the complexity of any computer's hundreds of gigabytes of electronic RAM.

It does not stop there. Goggles with special lenses can be (and have been for scientific tests) fitted before a subject's eyes. These lenses will invert the image the eyes pick up so that a right-side-up image is sent to the brain. Out of natural and repetitive function the brain instantly inverts any image coming along the optic nerve. Therefore the result is that the subject of the test is seeing everything upside-down. It takes some 2 weeks of enduring this terrible non-equilibrium before the subject begins to adjust to walking or doing most anything, for that matter, while seeing everything upside-down. Around this time something phenomenal happens. The brain has realized this is not transient and that it is not, more phenomenally, practical. It therefore inverts the image again, and even while wearing the special goggles the subject can now see right-side-up again.

This has nothing to do with our own will. Indeed, we didn't even know the brain understood such things on its own and took the appropriate counter functions to make sure proper sight would always be maintained without such inconveniences. It understands right-side-up, the center of gravity, and that there can be no black blotch in one's vision.

Even more mysterious and complex than the process of sight is that of speech. Speech expresses logic and reason, concepts and ideas. Sight takes in information, but speech is the expression of what has been taken in, learned and analyzed. It is therefore in its way the vital sign of the brain. The brain being a stupefying electronic machine, speech is itself thus dependent on steady electromagnetic rhythm.

Modern studies into the "body electric" have revealed not only what a remarkable electromagnetic machine the body is, but it has also revealed to what extent the body's biological functions are tied into the Earth's own magnetic field and electromagnetic waves.

Physicist Dr. Thomas Valone, Director of Integrity Research Institute in Washington DC, notes in his scientific monograph "EM Fields and Life Processes" that the alpha rhythm of the human brain is the same frequency as the Earth's fundamental resonant frequency (7.8 Hz.)

One tragic connection is found between crib death and solar storms. Solar storms are capable of compressing or decompressing the Earth's magnetic field. Researchers have confirmed that a higher number of crib deaths occur during these solar storms. This mysterious and tragic occurrence— the sudden death of babies who merely stopped breathing— now has some solution. It was known by studying previous victims that they had low melatonin in their blood. This substance controls the amount of nitrous oxide which is a compound essential for breathing. Theorists proposed putting the connection of magnetism to the test. They subjected baby rats to low level magnetic fields and indeed their melatonin decreased before they stopped breathing. This is an incredible advance for determining and hopefully in future preventing crib death or SIDS (Sudden Instant Death Syndrome).

A 1997 University of Massachusetts Medical School study also discovered a correlation between falling barometric pressure and the onset of labor— just another example of the relationship of the body to a steady and harmonious geo-environment. The examples continue, of course, but more to the point here is the possibility of a pulse or surge in the Earth's own electromagnetic field disturbing the speech centers of the brain.

Albert Budden has spent his life studying and writing about electromagnetic phenomenon, aberrations, and anomalies. One particular manifestation of electromagnetic aberration is "poltergeist effect." Rather than being viewed as some supernatural menace *a la* its German namesake "noisy ghost," it is better explained as an aberration or upset in a local area of the magnetic field. He observes:

> Television sets switch themselves on and off, repeated telephone connections are made which engineers consider 'impossible,' and computers show programmes that have not been installed by anyone or information that is inaccessible through normal use.

Such effects on electrical equipment and computers are the most intriguing since it naturally indicates a certain potential exists for electromagnetic influence on the human brain, itself a similar though vastly more complex electrical device. At the very least it seems electromagnetism can cause most anything in any device subject to its effects, including the implanting of information not personally experienced and altering the reception of information. If these are but the effects of a mild upset of the magnetic field in a local place, what would a massive flux or surge be capable of doing?

Could a major shift of this planet on its axis provide such a pulse? Since the Earth rotates at its equator around 1,000 mph and hurls through space revolving around the sun at around 68,000 mph, the rotational changes would wreak havoc on the core of the planet where the magnetic field is generated. Based on studies in dipoles of magnetic rock, it is established that the North Magnetic Pole was, amazingly, once located near Yucatan, it is speculated the result of some prehistoric "flip-flopping" of the field.

Theoretically then we have found an origin for the cause for a surge or pulse and for an effect on the brain. All are clearly tied in to the cataclysm of continental upheaval, memories of impacts, a great earthquake, cities falling and language suddenly being confused.

Speech it seems can indeed be suddenly affected by electromagnetic interference. The legend that such a thing once happened is hardly the stuff of fairly tales, especially when it is conveyed by peoples who had no knowledge of the body's complex electromagnetic functions; even more so when it goes hand-in-hand with stories of a giant earthquake and continental upheaval and veiled references to a moment of speciation and the division of peoples by new ethnicities and tribes.

For as equally sensitive as is the brain to electromagnetic rhythm so is that touchy world of DNA. And no two greater processes control the body: all major functions from the brain and all automated replication of cells from the DNA. The difference between the two is that the effect on DNA in reproductive cells would not show up immediately. It would only be evident in the next generation.

It is the potential of this vast invisible world that the ancients say was briefly revealed by a massive cataclysm involving an earthquake.

If it is true that a cataclysm suddenly confused languages— implying electromagnetic upheaval— what genetic effects would show up in the next generations? Could not many variations come about? Over generations as radically different types of people continued to be born, people would congregate according to appearance and language. Tribes and "ethnicities" would be born.

About one thousand years after the cataclysmic events described in the Bible, an interesting birth is recorded, with a smattering of surprise. It concerns the birth of Esau, the son of Isaac. Jacob, the twin of Esau, was a "smooth, plain man," whereas Esau came out "red all over like an hairy garment." He grew up to be an incredibly strong man. He also remained a completely hairy man all his life, so much so that when Jacob cheated Esau out of his blessing by pretending to be him before his blind father, Isaac, he had to cover his arms and neck with the hide of a goat in case Isaac should ask to touch him. In later life Esau was called Edom, meaning "red"— an interesting variation, almost our defunct image of a Neanderthal, to come out of two otherwise normal people who also had a very normal son like Jacob.

Births of such people were apparently not a biblical oddity or an exaggeration to make an ulterior point of Esau's more wild character. They had by Esau's time merely become a rarity. Ctesias, the Greek physician to Artaxerxes II, like all erudite Greeks, wrote a book when returning from Persia. His *Myriobiblon* (Book of Wonders) contained references to entire tribes of wild men in the mountains of India called by the Indians Calystrians, which means "dog-headed." Taking out some of the embellishments (like they had a tail), it is clear that Ctesias is speaking about some hairy people who lived in the mountains. They had no language and their clothes was the skin of wild beasts. "They are black and very honest, like the rest of the Indians, with whom they trade; they understand the Indian language, but they cannot reply except by barking and making signs with their hands and fingers like deaf-mutes. . .They live on raw meat. Their population may reach 120,000."

Megasthenes, Seleucus Nicator's ambassador to King Chandra-gupta, had been on the Ganges River and when describing what he saw merely quotes Ctesias. Both Greeks accurately described other Indian ethnography, including dark pygmies with long hair, which seem unquestionably to be the Veddahs.

The Persians, it is said, trained strong hairy wild men to fight for them. Like in accounts of Neanderthals, such wild men were found in the mountains.

Ancient histories abound with references to very different peoples. Not just isolated freaks, but tribes of giants and beast men. Giants may exist in unbelievable legends, but they exist in manner-of-fact histories as well. Hebrew histories tell of the Anakim. Writing of Ammon, the Bible records "The Emims dwelt therein in times past, a people great, and many, and tall, as the Anakims; which were also accounted giants, as the Anakims; but the Moabites called them Em-ims." The Ammonites called them Zamzumims, and "giants dwelt therein in old times." In Numbers 13:33 Caleb returns from scouting the land and says: "And there we saw the giants, the sons of Anak, of the giants: and we were in our own sight as grasshoppers, and so were we in their sight." The Anakim remained in Gaza, and Og was the last, finding the position of giant quite to his liking as it gave him the kingdom of Bashan, apparently populated by average people like us. The Hebrews record that his bed was still kept like it was a museum display. "For only Og king of Bashan remained of the remnant of giants; behold, his bedstead was a bedstead of iron; is it not in Rabbath of the children of Ammon? nine cubits was the length thereof, and four cubits the breadth of it, after the cubit of a man [13 feet]."

Giants survived until David's time, though fewer in number. The Philistines, too, an average size people, recruited Goliath, who was one of the *Raphaim*, an entire tribe of giants. Ishbibenob also fought for the Philistines, "of the sons of the giant, the weight of whose spear was three hundred shekels of brass in weight," who sought to kill David for revenge on Goliath. Saph also was one of the giants at Gob; and there was one of the giants in Gath who, as the Israelites marveled in 2 Samuel 22:20, "was a man of great stature and on every hand six fingers, and on every foot six toes, four and twenty in number; and he also was born to the giant."

Strong, powerful people in more modern times have struck some shorter races as giants. Pitched battles with Americans and even Jean Lafitte's pirates finally wiped out the Karakawas of Texas by 1860. To the Pacific Northwest Indians they were the progenitors of the Saskahaua Georges. The Spanish chroniclers recorded that this tribe was extremely primitive but unusually strong. Cabeza De Vaca added that they were unlike any other Indians. The Karakawas were tall, he wrote, 6-foot being common. They had the most remarkable physical prowess. "They go naked in the most burning sun. . .in winter they go out in early dawn to take a bath. . . breaking the ice with their body." With their bows "aimed at a bear in the top of a tree. . . the arrow went through the body and was propelled 40 to 50 yards beyond; though impelled nearly 200 yards already, the arrows were driven to the feather in the alluvial bank."

With time and wars the strange extremes of the human race which seemed so plentiful in our earliest histories have disappeared, at least those extremes we thought threatening. Giants are gone. Strange, hairy strong men, born from normal people like Isaac and Rebecca, though they numbered in tribes, seems a thing of the past, too. Veddahs, midgets and pigmies remain, those smaller folk we have found less threatening. Six fingers and toes pop up once in a while, but it is a feature passed on more frequently by the Amish. Having a smaller gene pool, intermarriage has assured that trait remains at the top of their "genetic code."

So many other traits have disappeared into an undertow of the gene pool. Take, for example, the 3-toed horse. It was long thought to be an evolutionary precursor to the modern horse. But examples exist where a "normal" modern horse has given birth to just such a 3-toed horse. For the last 2,000 years examples of it have been regarded as a "monstrosity" or "marvel," as the Roman Valerius Maximus called them. Richard Owen, Darwin's great detractor, considered them "*monstra per defectum*," and believed that the "prehistoric" species of Hipparion, which these modern 3-toed horses so resembled, could be bred back. So far it has proven to be too rare to provide enough offspring (a large enough gene pool) to do so.

Hipparion and many other species represent a strange paradox to the orthodox tenets of gradualistic evolution. Supposedly "prehistoric" or precursor lineal species are represented in the "gene pool" of a surviving related species today. Today, the Post-Neo-Darwinist theories of lateral speciation and the science of genetics have given us the answer. Such species were never "lineal." They were merely lateral variations of the same species. In this view, Hipparion did not die out per se but became absorbed through interbreeding back within the greater population of a related species. Is it the same for man and other animals? Are the depths of the genetic code a storehouse of once remarkable variations of different and past strains?

Thus Neanderthals, those often-times red hairy strong people, giants, midgets, pigmies, are all within us, within 3 billion base pairs. Once in a while a phenomenon might happen and one is born, but a burst of that unexplained world around us might open up that wonderful realm of the body electric and produce enormous variation once again.

# CHAPTER 10

# Doppelganger Inversion

Whether or not Sasquatch, Dsonoqua, Skoocoom, Almas, Neanderthals, etc., represent the phenomenon of prehistoric catastrophism's effects on biologic life— the morphing of something we know into something we would find hard to believe— we are confronted with a fact: *Ameranthropoides loysi* shows us anthropoids akin to monkeys have existed. From all that's been compiled, this type seems the origin of all the "hominids" Bigfooters and crypto-zoologists have sought. They have long legs, are purely animal, but because of their long legs they have more of a human body proportion. Thus they are enveloped in the most tantalizing enigma and, often, folklore.

One tribe of Sasquatch Men might indeed be humans, but there is that other tribe, the most enigmatic and human-like of them all— the owner of the Ruby Creek Print. That humanoid foot belongs to something very manlike in proportion. When a primate estimated at 7 and a half feet tall is able to simply step over a 43 inch fence we know it is proportioned like a human. It is not surprising therefore that it should be confused as a human or human monster. And indeed Jeannie Chapman called it a tall hairy man.

But beyond a small head, human proportion and long, narrow foot, we do not know what to make of it. Brave Peter Williams said of his pursuing Sasquatch: "Except that he was covered with hair and twice the bulk of the average man, there was nothing to distinguish him from the rest of us." Yet he describes that long, narrow inhuman foot.

Doppelganger inversion— two entities being confused for the other— is what makes it so difficult to excise the genuine Sasquatch from the Sasquatch Man. Another report, more disturbing than the Sacket Harbor report, tells us why. The account is in the *London Times* for January 4, 1784. The capture of a very bizarre human is documented:

> There is lately arrived in France from America, a wildman, who was caught in the woods, 200 miles back from the Lake of the Woods, by a party of Indians; they had seen him several times, but he was so swift of foot that they could by no means get up to him. He is near 7 feet high, covered with hair, but has little appearance of understanding and is remarkably sullen and subdued. When he was taken, half a bear was found lying by him, whom he had just killed.

Nothing more is heard of this strong wild man, and nothing more is made of him. But today we can make more of him in light of the behavior of the Dagestan Man. They are uncannily alike, suggesting that these wild men cannot stand any degree of confinement. Lake of the Woods is in Manitoba, far from the Saskahaua of British Columbia. Yet this hairy wild man was exactly what the Indians of the Chehalis Valley told J.W. Burns one tribe of the Sasquatch was.

The Manitoba wild man is not a novelty. In fact, doppelganger inversion is very frequently intrinsic in old reports of the 'wild man of the woods' across the whole continent. It is even intimated in one of the oldest reports in America: the 1818 Sacket Harbor incident where the creature was definitely an animal and yet it was described as "an animal that resembled the 'wild man of the woods.' "

There is also the story told by Otto Ernest Rayburn in his 1941 book *Ozark County* about the "Giant of the Hills." This hairy wild man was also about 7 feet tall and lived in the caves by the Saline River in the Ouachita Mountains in Arkansas. Since so many were afraid of him, sometime in the late 1860s the locals decided to finally capture him and put him in Benton jail. There they put clothes on him and tried to civilize him all to no avail; he ripped them off and escaped. Supposedly, he was recaptured and then there is no more heard of him.

Another wild man "of gigantic stature" was so well known to live

in the mountains of Arkansas since 1834 that peoples of three counties (St. Francis, Greene and Poinsett) knew of his existence. It was
in 1851, however, when two hunters in Greene County finally got a
clear view of him while he was chasing cattle. It was "an animal
bearing the unmistakable likeness of humanity," they reported. "He
was of gigantic stature, the body being covered with hair and the
head with long locks that fairly enveloped the neck and shoulders."
The only explanation locals could offer was the same given by those
hunters in Oregon in 1884; there they had thought the wild man
eating a raw dear was John Mackentire. In Arkansas they thought
this man a survivor of the 1811 earthquake that had devastated the
area; they thought he must have grown a body coat of hair in the
wilds.

A similar description is given for a 'wild man' seen near Warner
Ranch, California, in February 1876, by another hunter. "Looking
up he saw 'something' sitting on a large boulder, about fifteen or
twenty paces from him," recalled his companion. "He supposed it to
be some kind of an animal, and immediately came down on it with
his needle gun. The object instantly rose to its feet and proved to be
a man. This man appeared to be covered all over with coarse black
hair, seemingly two or three inches long, like the hair of a bear; his
beard and the hair of his head were long and thick; he was a man of
about medium size, and rather fine features— not at all like those of
an Indian, but more like an American or Spaniard."

There are more stories, of course. There is truth and error and
some reportorial gin in them, but one fact emerges: a very similar
looking hairy 'wild man of the woods' was a reality over the continent amongst completely unconnected Indians and Whites. Thus regardless of the fact that these were far from the Saskahaua there
seems little reason to doubt that they were technically the same
thing as a hairy "Sasquatch man." The Saskahaua was different in
that it was the last bastion of the tribe. Why not then believe the Indian stories? Why not then believe the Indian artwork showing wild
*human bukwas?*

Perhaps the doppelganger inversion makes it too controversial.
To accept that one tribe of Sasquatches was human is to except that
the manlike "animal human" is remarkably similar, both in size and
proportion. For the devotees of Bigfootery this axes the cone-headed
9 foot tall Bigfoot. But an animal-like man and a man-like animal
with longer hair on its head are far more fascinating concepts, far

above and beyond any modern theory of a Eurasian bipedal ape.[39]

But however distorted the "giant ape" Bigfoot became, an "animal" version of the wild man has at least had its advocates, loud and clear. Unfortunately, the only voice for humans was J.W. Burns, and information was already scant by his time, with the Indians already willing to accept anything hairy to be a Sasquatch man. Thus his writings amply reflect the enigma created by Indian convolution. And as such there is actually more truth in them than the later legend which wiped out humans and a very human-like primate and replaced it with a garish 9 foot Yeti *Gigantopithecus*. Not only does the convolution suggest people were still involved, it also reveals to us just how human-like the "ape" Sasquatch really is.

This can be especially found in Burns' significant January 1940 article in *Wide World* Magazine, a British periodical serving the entire British Empire. The article rated the cover art and a special promo at the bottom of the cover: "The 'Hairy Giants' of British Columbia." The double portrait painted between animal and human even prompted the publishers to place a cautionary preamble in which they call the Sasquatch both 'primitive creatures' and 'wild men.'

In the story of the kidnap of Indian maiden Serephine Long (who is pictured in the article) she made clear, as Charley Victor had, that Sasquatch men speak something like Douglas; and in her case she also said they had the ability of fire. Burns reports other seemingly human traits, such as that Sasquatches had a tendency to throw rocks as defense or offence, and even group together. This does indeed sound human, too; but we know the Skoocoom could do the same, and they were by no means human. Thus it remains enigmatic yet again exactly what Caulfield Anderson and his men truly encountered when they were held back and besieged at Harrison Hills in 1846.

Indians and old journals were not the sole source of Burns' information. One white man's story describes a 'wild giant'. Burns met him in 1937, and this was the first time he met a white man who had seen a Sasquatch. The encounter was only two weeks old. Roy King, a young mining engineer, was doing some prospecting by a stream in the mountains overlooking Harrison Lake (where Anderson had been in 1846). This stream churns its way through rocky natural walls as it flows toward the lake. It was in the shadow of these he pitched his camp. One day while strolling the rocky heights over the

---

[39] Some species of monkeys actually have longer hair on their head just like humans.

river, and coming fast onto his camp, he spied something surprising. "Thinking it was probably a thieving grizzly bear, King stopped and unslung both his rifle and binoculars. Focusing the powerful glasses he was startled by the image they brought clear and close to his eyes— a giant of a man entirely naked and excepting for a small space around the eyes, covered from head to foot with black fuzzy hair. The monster was interestedly examining the prospector's personal belongings. The young man admitted that at first he thought he had been too long alone in the wilderness and that he was 'seeing things.' Then it slowly dawned upon him that through the glasses he was actually getting a 'close-up' of the supposedly mythical 'Sasquatch.' " King waited until the giant strode off into the forest before he went to his camp and examined his things. Although tossed about, nothing was taken.

Chehalis Phillip was one of the Indians who also reported an enigmatic encounter with a rock flinging Sasquatch. His canoe was softly gliding down a tributary of the Chehalis called Morris Creek. When he was under the rocky precipice a boulder came hurling through the air and almost swamped his canoe. "Startled, Phillip hurriedly glanced upward and observed a huge man covered with hair leaping down the steep declivity with the agility of a panther," wrote Burns. "Under one arm he carried a bulky object that proved to be another boulder. Reaching a point of vantage, the giant deliberately slung the big stone straight at the now thoroughly-scared Phillip, missing the canoe by inches."

Was this a human or an animal? An even older account provides a sobering mug shot of something very human yet animal. This is the famous case of "Jacko." This was the name given by British Columbian citizens who captured a young "what is it?" near Yale, British Columbia, on July 3, 1884. It is recorded in the *Daily Colonist.*

## WHAT IS IT?
### A STRANGE CREATURE CAPTURED ABOVE YALE
### A British Columbia Gorilla
### (Correspondence of the Colonist)
### Yale, B.C. July 3rd

In the immediate vicinity of No. 4 tunnel are bluffs of rocks which have hitherto been unsurmountable, but on Monday morning last were successfully scaled by Mr. Onderdonk's employes [sic] on the regular train from Lytton. Assisted by Mr. Costerton, the British Columbia Express Company's messenger, and a number of gentlemen from Lytton and points east of that place who, after considerable trouble and perilous climbing, succeeded in capturing a creature which may truly be called half man and half beast. "Jacko," as the creature has been called by his capturers, is something of the gorilla type standing about four feet seven inches in height and weighing 127 pounds. He has long, black, strong hair and resembles a human being with one exception, his entire body, excepting his hands, (or paws) and feet are covered with glossy hair about one inch long. His forearm is much longer than a man's forearm, and he possesses extraordinary strength, as he will take hold of a stick and break it by wrenching or twisting it, which no man living could break in the same way. Since his capture he is very reticent, only occasionally uttering a noise which is half bark and half growl. He is, however, becoming daily more attached to his keeper, Mr. George Tilbury, of this place, who proposes shortly starting for London, England, to exhibit him. His favorite food so far is berries, and he drinks fresh milk with evident relish. By advice of Dr. Hannington raw meats have been withheld from Jacko, as the doctor thinks it would have a tendency to make him savage.

The mode of capture was as follows: Ned Austin, the engineer, on coming in sight of the bluff at the eastern end of the No. 4 tunnel saw what he supposed to be a man lying asleep in close proximity to the track, and as quick as thought blew the signal to apply the brakes. The brakes were instantly applied, and in a few seconds the train was brought to a standstill. At this moment the supposed man sprang up, and uttering a sharp quick bark began to climb the steep bluff. Conductor R. J. Craig and Express Messenger Costerton, followed by the baggage man and brakesmen, jumped from the train and knowing they were some 20 minutes ahead of time, immediately gave chase.

After five minutes and perilous climbing the then-supposed demented Indian was corralled on a projecting shelf of rocks where he could neither ascend nor descend. The query now was how to capture him alive, which was quickly decided by Mr. Craig, who crawled on his hands and knees until he was about 40 feet above the creature. Taking a small piece of loose rock he let it fall and it had the desired effect of rendering poor Jacko incapable of resistance for a time at least. The bell rope was then brought up and Jacko was now lowered to terra firma. After firmly binding him and placing him in the luggage car "off brakes" was sounded and the train started for Yale. At the station a large crowd that heard of the capture by telephone from Spuzzum Flat were assembled, each one anxious to have the first look at the monstrosity, but they were disappointed, as Jacko had been taken off at the machine shops and placed in charge of his present keeper.

The question naturally arises, how came the creature where it was first seen by Mr. Austin? From bruises about its head and body, and apparent soreness since its capture, it is supposed that Jacko ventured too near the edge of the bluff, slipped, fell and lay where found until the sound of the rushing train aroused him. Mr. Thos. White and Mr. Gouin, C.B.E., as well as Mr. Major, who kept a small store about half a mile west of the tunnel during the past two years, have mentioned having seen a curious creature at different points between camps 13 and 17, but no attention was paid to the remark as people came to the conclusion that they had either seen a bear or a stray Indian dog. Who can unravel the mystery that now surrounds Jacko? Does he belong to a species hitherto unknown in this part of the continent, or is he really what the train men first thought he was, a crazy Indian?

It is amazing that the article is introduced as "what is it?" plus "strange creature" and then "British Columbia gorilla." The consternation these "wild men" inspire is remarkable.

No one knows whatever happened to poor "Jacko." He either died or escaped or was let go. The long hair, the humanity of the body proportions, are those of genuine wild men, but Jacko was clearly an unknown type of primate. Or is the story all bunkum, as some have suggested? If so, why is Jacko a disturbing mirror of the description of the *orang pendek* of Sumatra and Borneo, although no one in British Columbia at the time had ever heard of such a legend?

Although Jacko is very different from what William Roe reported, it is not necessarily at odds with it. The long hair may only apply to males; another possibility is that Roe saw a very aged female that was losing her hair.

But one thing they have in common. Each is far too human in appearance to have been an Almas or *loysi*. Such an ape foot as both have could never pass as human, a crazed Indian or anything manlike. The Ruby Creek Print is a reminder that something very humanoid also walks the Pacific Northwest.

Although I speak disparagingly of Albert Ostman's account of his kidnapping, one wonders if he did not see real footprints in his many prospecting wanderings and upon these built his fanciful account. I ask this here because he describes the Sasquatch foot as having a huge big toe, even exaggerating to say that's all the Sasquatch needed to climb up a mountain. Jacko is a remarkable mirror of the *orang pendek*. That footprint Jacobson traced at Danu Bento swamp is a disturbing mirror of Ostman's account. Is this the actual very-human Sasquatch? It is a giant similar to the *pendek*? Is it something like Man that came from the Old World?

It should not be surprising or even ironic therefore that Siberia holds the final part of the answer to the enigma of the Saskahaua. It holds a special paradox to its reputation as a place of utter desolation we associate with human banishment. Siberia teems with wild-life. It is dotted with lakes and etched with rivers. To banish one here is like banishing them to the Pacific Northwest and the plains of the far north. It is a beautiful place of taiga forests and flat, other-worldly tundra. Indigenous peoples build and inhabit quaint log cabins, and the bells of their reindeer herds tinkle through the for-ests. Human habitation is still so unsullied by the outside world that stories of two tribes of men remain intact and aboriginal. To go to Siberia is to go back in time in America to a more innocent and reli-able age in reporting the existence of what we seek in the Saskahaua.

Reindeer herders have maintained that in the mountains and for-ests of eastern Siberia there lives a large hairy creature called Mirygdy, meaning "broad shoulders." This characteristic is accentu-ated by the fact it has no neck. It is very tall and definitely not hu-man. It is very aloof. It eats raw meat like a primate, tearing it and putting it in its mouth rather than gnawing at it off the carcass like wolves. Since it is seen over such a vast territory as eastern Siberia (Chuckchi, Lamut and Chukotka territory) to as far as *western* Sibe-ria (5,000,000 square miles), there are a variety of names for it. Each in its way helps to describe it. In the Komi Republic of western Si-beria it is known as Tungu. Toward the east a similar type (or varia-tion) is called "goggle eye" (Kiltanya) because of its large eyes. It is also called Teryk (Dawn Man), Girkychavyl'in (swift runner) and, in-terestingly, Rekhem or Dzhulin (sharphead). The Chukot Peninsula is particularly noted for sightings of this mysterious creature. This peninsula is the remaining area of the Bering Land Bridge over which conceivably several of the species could have migrated into North America in the ancient past.

For the sake of clarity and consistency I will refer to "it" from now on as the Dawn Man. The similarities between the Dawn Man and the humanoid "what is it?" Sasquatch of the Pacific Northwest is obvious. It is so obvious that to describe Dawn Man is to merely describe the very human Sasquatch (Ruby Creek, Jacko). He is about 6.6 feet high, sometimes closer to 7 feet, and his face is man-like with high, protruding cheekbones, the bridge of his nose narrow. Dawn Man is also noted for a high pitched inhuman call. He has hair all over his body, but the hair is longer on the head, like a man. He is called Dawn Man because that is usually when he is seen. He is largely nocturnal. Dogs are unusually frightened of him. A de-scription of Dawn Men feet (in this case called *tungu*) is given by Myra Shackley and it confirms that Ruby Creek Print hitherto bur-ied in American reports.

The *tungus* are always described as very tall (about 6 ft 6 in.), which may not seem excessively tall to us but both Tungus [actual human tribe in the area] and Samoyeds are rather short races. The *tungus* live in the coniferous forests, supposedly beneath projecting tree roots (a belief which finds an echo in medieval Wildman legends) and cannot speak, although they can whistle, make inarticulate cries and call something like 'ru-ru-ru-'. In 1962 one *tungu* threw sand into a tent of a herdsman and made his characteristic cry— but no one went out, as they were afraid. The herdsman in question, Evgenii Grigor'evich Tog, had seen *tungu* tracks before, in 1929, and described them as being rather long and narrow.

Children surveyed in the Komi Republic noted that the *tungu* was always associated with a high, piercing howl. This survey was initiated by Soviet authorities in the early 1980s of the pupils of the vocational schools of Salekhard, the principle town on the Ob River.

In addition to the many physical similarities to the Sasquatch, the majority of Dawn Man sightings occur in autumn, October being the prime month.

Can it just be coincidence? The artwork of the Pacific Northwest Indians comes into play again. There is that rare variation in the *bukwas* masks that cannot possibly be the *Ameranthropoides loysi* or humans. It has that strange nose with a bridge that curves down to the upper lip. A dance mask of the Tsimishian tribe is one of the best examples (and frequently used to promote Bigfoot) of an unknown *Old World* anthropoid having had contact with the Indians of the Pacific Northwest. Aside from the typical *bukwas* features of a snarling mouth, baring teeth, it has the beady, deep set eyes. Yet it is not human and the prominent ocular socket of a *loysi* is absent. The teeth are completely even, with no hint of canines. Along with the very human nose (Catarrhinian), the mask reminds one of General Topil'skiy's description of that dead Almas in the High Vansk, and the Sasquatch that William Roe saw on Mica Mountain in 1955.

Indian art varies, of course. There were good artists and bad artists, just like in any culture. There were some who also no doubt got a better look at what they later whittled into wood and those that did not. But Indian artwork does not show fantasy. It shoes symbolism, stylization, and at times frightening accuracy.

And if the wild men lived up to their name, it is not likely that any Indian frequently got too close. The consistency of the *loysi* features in Indian artwork is remarkable. Therefore the rarity of the Tsimishian mask suggests it represents the most violent of the wild men, the ones who were held in the greatest dread. Its stark variation

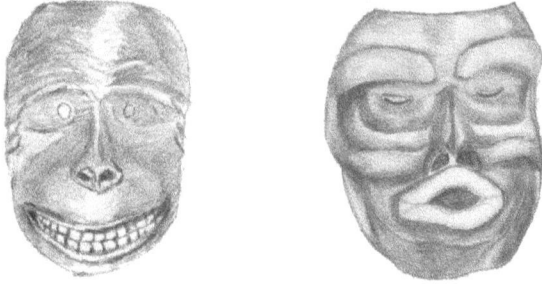

*Tsimishian Mask compared to a standard Dsonoqua mask. The features of the first highlight the violent nature of the wild man of the woods whereas the Dsonoqua mask underscores the howling and whistling nature of the nude tribe in the forest.*

from all the others also demands our attention, especially when that stark variation is consistent with the descriptions of both Old and New World witnesses who saw a living example of the same thing.

Is the Tsimishian Mask the Dawn Man? If so, what does the Sasquatch truly look like? Sightings of similar Dawn Men may tell us, and amongst these there is a disquietingly human description. Sightings have not been limited to Siberia. The oldest reports have actually come from the primeval forest of Shennongjia in central China, and at the same time of year of October. The oldest reliable modern report comes from a biologist, Wang Selin. He describes a face similar to the Tsimishian Mask. The incident happened in World War II, in September or October 1940. While driving to Tianshui from Baoji he and his traveling comrades heard gunshots ring out ahead in the road.

> . . .When the car reached the crowd that surrounded the gunman, all of us got down to satisfy our curiosity. We could see that the 'wild man' was already shot dead and laid on the roadside. The body was still supple and the stature very tall, approximately 2 meters. The whole body was covered with a coat of thick grayish-red hair which was very dense and approximately one *cun* [a little over an inch] long. Since it was lying face down, the more inquisitive of the passengers turned the body over to have a better look. It turned out to be a mother with a large pair of breasts, nipples being very red as if it had recently given birth. The hair on the face was shorter. The face was narrow with deep-set eyes, while the cheekbones and lips jutted out. The scalp hair was roughly one *chi* [aprox. 1 foot] long and untidy. The appearance was very similar to the plaster model of a female Peking Man. However, its hair seemed to be longer and thicker than that of the ape-man model. It was ugly because of the protruding lips.

According to the locals, there were two of them, probably one male and the other female. They had been in that area for over a month. The 'wildmen' had great strength, frequently stood erect and were very tall. They were brisk in walking and could move as rapidly uphill as on the plain. As such, ordinary folks could not catch up with them. They did not have a language and could only howl.

A fleeting sighting occurred in China on May 14, 1976. Six cadres driving through the forest saw a tall, tailless creature covered in reddish hair. The driver cast the spotlight on it. All got a good look at it, enough to realize they had just seen something unheard of. A telegram to the Academy of Science sparked a massive worldwide response.

Ironically, the Chinese met with the same dismal results the Russians had after Pronin's 1958 sighting. An army consisting of 100 scientists and trained personnel armed with tranquilizer guns, tape recorders, cameras, and hunting dogs, assisted by army scouts and local commune members, combed the area for the better part of 2 years. Unlike the Russians, however, the Chinese did come across hairs, feces and footprints, but the hairs proved only to *not* be from bears. As other primates are known to live in the forest this was of little benefit.

However, Chinese histories are particularly rich with enigmatic references. Qu Yuan, the statesman-poet of the Warring States Period of China (c. 340-278 BC), in verse mentioned the existence of "mountain ogres" in the Shennongjia. Historian Li Yanshou of the Tang Dynasty (618-907 AD) describes the existence of hairy men in another county in the same province as the Shennongjia (Hubei). And Yuan Mei, the poet of the Ch'ing Dynasty (1716-1798) in his book *New Rhythms* speaks of a creature which is 'monkey-like, yet not a monkey' existing in Shaanxi Province.

The enigma of the Shennongjia has continued until the present. On June 30, 2003, there were 4 eyewitness reports of the Yehren, as the Chinese call it. In each report he was reported at around 5.5 feet tall, grayish skin, covered with black hair; that on his head being shoulder length. His back was "crooked" and so were his arms. His facial features, unfortunately, remain nebulous. He was seen quickly crossing a road and on the edge of the road witnesses later found footprints about 12 inches in length and then a 3 meter long patch of what appeared to be urine. Fortunately one of the witnesses was a Chinese reporter, Shang Zhenming. Thus the report quickly found its way in the *China Daily* and from there to the West.

[ 223 ]

In the subsequent news feeding on the "Chinese Bigfoot," the popular mistake was sadly reinforced: "'Bigfoot' is more commonly known as the 'Yeti,' a mythical beast partly human and part animal has remained an unsolved mystery of the Himalayas."

China has its own paleoanthropologists in the pursuit. Among them are Yuan Zhenxin and Zhou Gouxing. Zhenxin, now retired, displayed a footprint cast— those ever-so tangible pieces of evidence— retrieved from one of the earlier scientific expeditions into the Shennongjia. Like always, these ever-so-indispensable bits of evidence only serve to confuse the issue. The print is a flat large human type print with absolutely no connection or similarity to the famous Yeti, Dawn Man or Sasquatch.

Ambiguity as a consequence will have its unavoidable ill effects here, too. The Chinese came to regard their "Yeti" as *Giganto-pithecus* or something close. Paradoxically, the obscure details of the Yeti have allowed its muddled image to be superimposed on the report of almost every "hairy hominid." And for some reason, with lack of clear proof, everybody has reached at a few ancient fossilized giant teeth and fingered them as the culprit. This makes the Himalayan Yeti *Gigantopithecus*, the American Bigfoot *Gigantopithecus*, the Canadian Sasquatch *Gigantopithecus* and the Chinese Yehren *Gigan-topithecus* in spite of the fact all have such different feet.

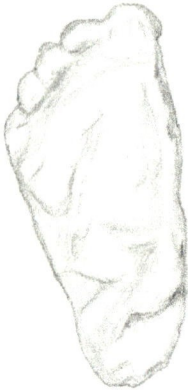

Yehren Print

"A bird in the hand is worth two in the bush" may be cliché, even trite, I suppose, but in the end all clichés come true. The pattern is obvious. We have reached, however incongruously, at things we know to explain Sasquatch, Almas, Teryk, and Yeti. We have giant ape-like teeth and that is enough. But has anybody considered we have something entirely unknown in Sasquatch or Teryk or Almas? Who had suspected a *loysi* existed until de Loys shot one? Nobody suspected an anthropoid 'version' of a known genus of monkey. No one has fossil bones from such a *loysi* either. Yet it existed.

But Yuan Zhenxin's Yehren footprint is not entirely at odds with the true enigma. It reintroduces people into the picture again. Although the voice in America is silent and in China confused, in far eastern Siberia and Yakutia just such encounters with hairy wild men underscore the Amerindian traditions of at least one tribe of human footprints. In Siberia such people were called Mulon, in

Yakutia they were called Chuchuna. These were the very primitive people that P.L. Dravert tried to get the world to acknowledge so long ago. Echoing through the foggy history of these people are rumors of holocaust and secret mass graves. According to Myra Shackley "Chuchuna-hunts were organized during Tsarist times and during the Second World War, resulting in the deaths of many, and secret burial of the corpses." During the Russian Civil War, it is said that many were also killed. Thousands of people were moving into previously unoccupied land in the area, and it is thought that the Chuchuna's violent nature necessitated killing many of them, first in self-defense and then in organized parties.

The land of Yakut, that area of Russia north of Mongolia and China, is the perfect place for such people. Most of Yakutia remains a heavy forested, rugged and very cold land, little populated or explored. Even many of the mountain heights remain mysteries. There is little more than reindeer herders inhabiting the wistful solitudes, and the towns remain largely grouped along the Lena River. Even into the 20th century it was once standard practice among the native peoples when traveling out into the forests to take precautions against the Chuchuna. The reason is more appreciated from the Chuchuna's other name of Mulen, which seems to mean "thief." The Mulen were credited as attacking hunters and any type of expedition.

According to the locals of the areas of Siberia and Yakutia, the Mulen and Chuchuna were known to wear clothes and to hunt with bows. There is often disagreement on their height, but most descriptions of them in general echo what the Amerindians tried to make clear was their wild man of the Saskahaua. One witness, Tatiana Il'inichina Zakharova, said of her encounter:

After the revolution, in the 1920s, inhabitants of our village met a Chuchuna, while gathering berries. He too was plucking berries with both hands and stuffing them in his mouth, and when he saw the people he stood up straight. He was very tall and lean, say over 2 meters. He was dressed in a deer skin, and was barefoot. He had very long arms and disheveled hair on his head. He had a big face, like a man's but dark. His forehead was small and hung over his eyes like a peaked cap. He had a big chin, broad and much bigger than a man's. All in all he was like a man, but of much greater stature. After a second he ran off. He ran very quickly, leaping high after every third step.

From descriptions he heard, Dravert painted a picture of a smaller man. In his 1936 work *Dikie lyndi muleny i chuchuna* (The Wild Mulen and Chuchuna People), Dravert describes them more like a Neanderthal:

> He is less than the average man in height. His hair is unkempt, and the greater part of his face is covered in hair. He wears animal skins with the pelt outside. He wears something similar to deerskin boots and a band is tied around his head as a headdress. Sometimes the clothing is of undressed hide, as with the coat, breeches and boots taken from a Mulen by the Tungus who had killed him, perhaps five or six years ago. For armament he has a bow with about 100 feathered arrows with special heads of unknown construction. He carries them over his shoulder in a special sheath or bag. The knife hangs from his belt, a metal one, like those produced by wild uncivilized peoples. He has fire-steel. Sometimes he has a spear with a head of unknown construction. Often he is met with armed only with sticks and stones. He produces separate inarticulate noises. Stealing up to a travelers' camp or to the tents of the local inhabitants the Mulen shoots with his bow or throws stones. He is in the habit of shooting non-stop until he has exhausted all his arrows. When he has run out of arrows he flees. He shoots from 15 to 20 sazhen [up to 47 yards]. They are supposed to live in caves.

Knowledge of the Mulen and Chuchuna in Yakutia and Siberia was once so broad that a local communist official finally mentioned them to Moscow. The official minutes of the Party Secretary A.N. Asatkin (from Vladimirskii) for the Commission on Yakut Affairs in 1928 shows the surprise he held that his superiors of the Central Committee of the Communist Party had taken no official interest in these strange people.

With the rumors of massacres and hunts it is possible the Central Committee didn't want to hear. To what extent the Mulen and Chuchuna fought back is also hard to say. In 1933 Dravert made an open plea that the hunting of them should stop, something not necessary if the killing of these wild men was purely in self defense. He went on to insist that they had the rights of any other citizens of the USSR. One can understand why Porshnev entitled his "report" *Soviet Ethnology*. He had relied heavily on Khakhlov and Dravert's accounts, and mixing them together he was sure these were some type of "subhuman."

Subhuman is a fun theory, and the world of Rice Burroughs is one that delights the imagination, but subhuman as a concept could only exist in the world of visual impressions molded by passé Gradualism's hopes. Neanderthal existed for the Russians as an option to explain all these reports for that very reason. Gradualism (then dominant) estimated Neanderthals were still alive 30,000 years ago, time enough to lose the ducks perhaps but not to wipe out a "species of men." Therefore Porshnev and some of the Russians and other anthropologists (e.g. Shackley) who did not mind treading into the popular forum of "wild men" and leaving their mark (for which we are all grateful) were able to speculate on Neanderthal Man's continued existence. But nobody, and I mean nobody, would have proposed that *Homo erectus*— Peking Man— was still alive. This would have gotten them all drummed out of the service. Yet today we now accept that *Homo erectus* (Java Man/Peking Man) was fully human. With Prof. Laughlin's work of U of Connecticut we know very similar skull types exist today, in that very part of the world. They were/are simply a very interesting variation of ourselves, *Homo sapiens.*

Peking Man's bones, interestingly, contribute to the world of doppelganger inversion. European savants reconstructed Peking Man to look like Mongloid apemen because they fancied these were the ancestors of "Chinamen." But what did they really look like? Could they be the Mulen and Chuchuna or what Tatiana Zakharova saw? Could they indeed be human, as more recent studies suggest *Homo erectus* was? But what did Wang Selin see? He said it was close to Peking Man reconstructs, but he wasn't looking at anything human.

Creationists are credited with grudgingly insisting that Peking Man was nothing more than an ape of sorts. Interestingly, they stand on a long and intermixed pedigree with evolutionists. Some of the early Darwinists to examine "Dragon Bone Hill" (Chou K'ou Tien in Chinese, the site of Peking Man finds) puzzled over why mostly only heads and jaws were found of Peking Man and why they were also intermixed amongst other animals. Frere Teilhard de Chardin even noted that there had been a fire pit that went down 6 meters. At his invitation an expert on the Stone Age, Professor Henri Breuil of the College of France and l'Institut de Palaeontologie Humaine, visited the site (1931) and had a free hand for 19 days. Breuil found surprising evidence that this was a center of some kind of industry. Stone chippings were 18 inches deep over the area, and stone tools had been imported from more than a mile away. Antler bones had been used to do work or for ceremonial reasons, and logically the deer they belonged to must have been used for food. The "traces of fire," as some had reported (to accentuate that these "early men" used crude campfires), was actually some kind of furnace operation. The ash heap went down 23 feet and the minerals in the surrounding stone had been fused, thus indicating the fire had been hot and kept burning for quite sometime.

This was hardly a campsite of roving, wild, primitive men. In another location (called "Upper Cave"), a few intact fully human skeletons had been found. Were these the actual masters of Dragon Bone Hill? Were the skulls of "Peking Man" their victims? Of the 40 or so individual skulls retrieved from the dig, each showed signs of having been broken open in the back in a very exact way, indicating decapitation and that the brains were removed. Monkey brains remain a delicacy in parts of Asia today, and Dr. A.C. Blanc (University of Rome) even noted that a tribe in New Guinea kills and removes monkey brains in the exact manner. The eminent Marcellin Boule

*Did "Peking Man" truly exist? And if so did he really have an ape-like muzzle like the Tsimishian Mask?*

himself traveled to the Dragon Bone Hill dig site, only to be upset that he traveled around the world to see remnants of what he considered monkey skulls. He noted all the evidence of industry in the area and that man had been in charge of the site. "We may therefore ask ourselves whether or not it is over-bold to consider *Sinanthropus* monarch of Chou K'ou Tien when he appears in its deposit only in the guise of a mere hunter's prey, on a par with the animals by which he is accompanied."

Is therefore Peking Man truly *Homo erectus* and therefore an ancient example of *Homo sapiens*? Or are the reconstructed skulls simply too humanized in the hands of hopeful Darwinian artists? Or is Dragon Bone Hill the site of cannibalism, one type of man eating another that was considered too animalistic or primitive? Or were the monarchs of Dragon Bone Hill eating something quite human in appearance but which was only ape, like the Almas or Dawn Man? What kind of truly different people can be out there? Only discovery will tell us. But just as some humans today are cannibals, so too was ancient man. Those who have no problem eating other people have no problem eating a man-like primate. Modern man may also have had no trouble "hunting" them, as the Tsarists and Bolsheviks hunted the "people" Dravert tried to protect.

There is more than enigma to the Mulen and Chuchuna; there is confusion that will never be cleared up due to the rumors of genocide. The same enigma will remain over the Saskahaua for different reasons. The humanity of the Dawn Man, though it is clearly not a human, will forever cause us to question just how much humans were behind the stories of the Sasquatch. If humans were part of the Saskahaua enigma, were they a depraved branch of the Mulen and Chuchuna? And were (are) these in turn the last strains of, of what? Ancient Neanderthals? Peking Man? Or something totally unique waiting to be discovered? In the end the confusion of doppelganger inversion wins out. We simply do not know what inspired all these similar reports over America and Siberia and China. But whatever it was, it seemed or was (both) very human.

CHAPTER 11

## Bone of Contention

America has provided us with a photograph of what we would never have expected: an anthropoid closely related to monkeys. But in contrast to the New World, the Old World has provided nothing concrete but a mass of anecdotal stories and a few plaster casts of footprints. In one area, however, it *has* given us something America has not. Fossil bones have been dug up that, given the reinterpretation of the sediments by Flood Geology, may prove to be the most tangible corroboration that something like the Almas and Dawn Man did and *does* exist, adapting and migrating.

The skeletal peculiarities of one have made it almost a household name— *Australopithecus.* They were so named "Southern Ape" in 1924 when first found in southern Africa by Raymond Dart. Their difference from apes was initially so marginal that they went unnoticed for about 35 years until 1958. In that year Dr. Louis Leakey discovered fossil anthropoid remains in sediment at Olduvai Gorge, Kenya. This was during a time when science was dominated by Uniformitarianism. Therefore Leakey calibrated the sediment thickness by the centimeter. The age of the fossils appeared by this method to be something like 2 million years old. The find caused excitement because Leakey then extravagantly declared his fossilized remains to be *Zinjanthropus boisei*— "East Africa Man"— and, moreover, to be a true early ape-man.

With the help of the likeminded *National Geographic Society,* which then backed Leakey's next dig and was looking for major news, Leakey was credited with making a momentous and unique discovery. As it turned out, however, after lengthy comparative studies, Leakey had to admit that what he found was a variety of Australopitheci. His *Zinjanthropus* was subsequently renamed *Australopithecus boisei.*

Instead of Dart's *Australopithecus* re-cataloging the Olduvai Gorge finds, ironically *Australopithecus* was regrouped as an early hominid. Once again with the help of *National Geographic* this once discredited "southern ape" was attaining worldwide attention as Leakey's ape-man. Contributing to the excitement was the fact that Leakey found tools nearby, fueling the furor by declaring that his East Africa Man was a toolmaker. Subsequent digs, by both Leakey and his son, Richard Leakey, however, only added confusion. In Bed II Louis found bones that matched *Homo erectus* and what he called *Homo habilis* (Handy Man), one clearly very close to modern humans. Thirteen years later, Richard added to the controversy when he discovered "bones virtually indistinguishable from modern man" deeper down in the sediment. Since Uniformitarianism's dating methods did not allow all to be living at the same time, the remains were thought to confirm true hybrid ape-man status for the Australopitheci.

The uniformitarian model, as always, was non correspondent in logic. How can *Australopithecus* be an ancestor of man if man, by the depth of the sediments, preceded it? Flood geology, however, provides a better answer: that the fossils represent the people and animals living at one time and buried at one time. We have then found the skeleton of something that lived more recently alongside mankind.

This deduction is strengthened by a find in Bed I, the lowest level of the dig, when Louis Leakey found remains of a circular stone hut, similar to the kind still made in Africa today, indicating that the remains are simply those of creatures all living together and buried in the sediments of a more recent and probably local deluge.

This only makes the Olduvai finds more exciting. Delving deeper into the details of *Australopithecus* uncovers striking similarities between it and the Almas. Australopitheci do seem to vary in size, but the largest known of them, the *Robustus* (Paranthropus), is about the size of a medium man, which fits the Almas size. It does have a

[ 231 ]

much, much flatter face than any of the large apes. Furthermore, it had a small cranium and *no prominent sagittal crest.* Its head only slopes back. So far, all this is in-keeping with the reports of Almas. Some of it, however, overlaps with reports of the long haired Sasquatch, which to an extent we should expect if the "genuine" Sasquatch (non-human, non-Skoocoom) is of the same general origin. Therefore I will on occasion note the similarities.

The most striking similarity to both is *Australopithecus'* teeth. Roe highlighted the straight, white teeth of the Sasquatch; Indian bukwas masks show the same teeth, and Almas witnesses speak of the same thing. Both Dart and Leakey were struck by how human *Australopithecus'* teeth were, only bigger. Raymond Dart noted the Australopitheci teeth were straight across and missing the long canines of apes. Leakey had even nicknamed his find "nutcracker man" because of its strong manlike teeth. The front teeth, incisors and canines, remain small like in a man, and the dental arcade (the curve of the jaw) is less U-shaped than any modern ape and more parabolic. Huge jaw also calls to mind the Almas, whose teeth can crush a squirrel's skull.

Placing an *Australopithecus Robustus* next to a modern skull of an orang reveals to what extent the muzzle must have been much shorter on australopithecines, thus more manlike. Its brain was small, about 500cc (cubic centimeters), one third that of a human.

Of all the australopithecine bones that have later been found, there appear to be two distinct kinds. The other is dubbed *Australopithecus africanus* because it was more gracile, with smaller teeth and bones; it is the size of a big chimp (sound like Jacko or the orang pendek?). But even this smaller version had molars that were larger than modern gorillas. Thus the jaws of both species are large and the cheekbones are high.

A brilliant anatomist, Solly Lord Zuckerman, and UCLA's Dr. Charles Oxnard, headed a 15 year long study of man, apes, monkeys, and australopithecines to try and come to a firm decision on

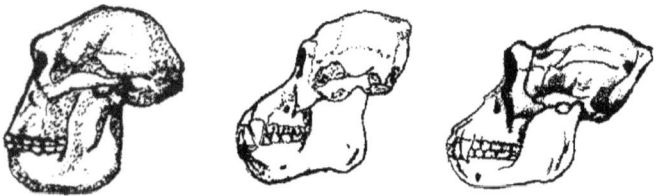

*Robustus (reconstruction) compared to an Orangutan and a Gorilla. Note the great difference in the muzzle area.*

their relationship. Although they did the most protracted comparative study to date, much of their work was eventually swept away by
the increasing desire to place *Australopithecus* as a completely upright human-footed missing link. But now that paleontology is leaving off australopithecines as anything but, many are coming back to
appreciate the depth of Zuckerman and Oxnard's work. What they
discovered is remarkable in light of our current study. Oxnard's observations are the most applicable here. He stated:

> Multivariate studies of several anatomical regions, shoulder, pelvis, ankle,
> foot, elbow, and hand are now available for the australopithecines. They suggest
> that the common view, that these fossils are similar to modern man or that on
> those occasions when they depart from the similarity to man they resemble the
> African great apes, may be incorrect. Most of the fossil fragments are in fact
> uniquely different from both man and man's nearest living genetic relatives, the
> chimpanzee and gorilla. To the extent that resemblances exist with living forms,
> they tend to be with the orangutan. . .Finally, the quite independent information
> from the fossil finds of more recent years seems to indicate absolutely that these
> australopithecines. . .are not on the human pathway.

Every time Zuckerman tried to confirm a relationship "I have
ended in failure," he would declare.

What Oxnard and Zuckerman are saying is that australopithecines are related to *nothing living today*, at least that which we have
classified and that of which we know.

Physiologically they are unique primates. They are neither man
nor ape, per se. This is poignantly found in the peculiar differences
in *Australopithecus'* ear ossicle. Modern man and the great apes are
even more similar to each other in their incus than to the incus in
the *Australopithecus robustus*. It is possible, therefore, that it did
walk very differently from both man and African apes (which are notorious for waddling before going quickly down on all fours).

Equally engrossing is the study by Jack T. Stern and Randall
Susman, both anatomists at the University of New York at Stony
Brook. They also did a protracted study, and published this in the
*American Journal of Physical Anthropology* on July 2, 1983. This
laudable study showed that *Australopithecus* was adept at tree
climbing, therefore explaining the long curved fingers and long
arms. Foot studies proved it was strong and heavily muscled, one
adept at climbing as well as possible bipedality. "There is no evidence that any extant primate has long, curved heavily muscled
hands and feet for any purpose other than to meet the demands of

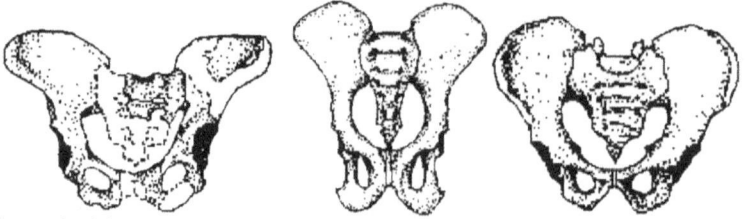

*Australopithecus pelvis, left, is contrasted with a Chimpanzee (center) and human (right). It is not some cross; it is different from both.*

full or part time arboreal life."

However, if they were indeed bipedal it is surprising that Susman and Stern found absences of certain traits found in humans. For the absence of or just a weak sacrotuberous ligament (needed for human bipedal locomotion) they offer: "One possible explanation is that the bipedal gait was like that of chimpanzees or spider monkeys." And then later they added significantly: "The possibility that the sacrotuberous ligament of *A. afarensis* [species of *africanus*] was not as powerfully developed as in humans suggests either a lesser frequency or a different manner of terrestrial bipedality than typifies modern humans." They also noted the obvious lack of identity between the sacrum of australopithecines and those of humans. Inasmuch as it was even more coronal than in chimpanzees, they comment: "It suggests to us that the mechanism of lateral pelvic balance during bipedalism was closer to that of apes than humans."

Examination of the femur, which was close to a human femur, revealed that the coverage of the femoral head was larger. This would suggest that the creature walked more like the ape family and was able to abduct the hip in that manner.

One would think that the foot of the australopithecines would be very relevant here to determine if it could make the Sasquatch footprint. Unfortunately, what has been found of them is hotly contested. Oxnard and his assistant F. Peter Lisowski note that bones of the Olduvai foot were manipulated incorrectly to give it a more human appearance. But when they placed them back in their proper places: "It is thus clear, a) that the Olduvai foot is not adapted for bipedality in the manner of man, and b) that it displays features in which it resembles the feet of arboreal creatures. Such anatomical characters as relate to bipedality in the fossil suggest usage as in an arboreal species that also walks bipedally with flattened arches (like a chimpanzee or gorilla) rather than the high arch of man."

Manipulation of the foot bones back into their original and correct position gives the australopithecines an offset big toe. It is not,

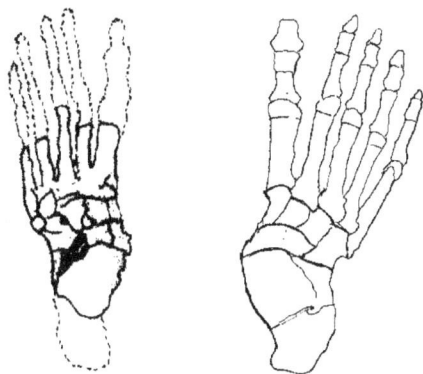

*Australopithecus foot. The length of the toes is subjective since the phalanges remain missing.*

however, like in other apes. It is not opposing in the typical manner. The upshot is the foot is flat with a big toe that is shorter and offset from the others. Although the bones of the foot remain incomplete, this fact is one of the most intriguing there is, as no ape-like creature has such a reported foot except the "manlike" Almas.

Everything about *Australopithecus* indicates a type of primate that is neither man, ape or monkey. It is unique and separate, distinct and yet similar to them all. It could indeed be the Almas or something similar. At the very least *Australopithecus* is useful for pointing to the possibility that a unique type of primate existed and perhaps still does. The Sasquatch and Skoocoom, even with their radically different feet, can hardly be considered imaginary like a pink dragon or hydra or some other mythological being. Their monkey-like, long feet and humanoid proportions are a strong link to monkeys and yet show they are separate and distinct from all other primates.

One of most impressive accounts of "Bigfoot" comes from a deputy sheriff, Thomas E. Dillon, badge 2007, of Clearwater County, Idaho. It is impressive because he describes the walk of the "Bigfoot," and what he reports offhand matches neither ape nor man but that which the scientists above speculated upon for *Australopithecines.*

While on duty on the night of October 25, 1979, he was approximately 1.2 miles over French Mountain Saddle. Report heading: "Sighting of an unknown type animal." His affidavit, given November 11 of that year, reads:

While on a routine patrol on French Mountain Road, at approximately 0300 hours, this officer did observe, heading towards Bungalow, it was a foggy night, slight rain mist falling. While on routine patrol I did observe out of my peripheral vision something which I determined to be an animal. I backed up, took a handheld 300,000 candlepower spotlight, put it on this thing, at the first sighting it was approximately 15 yards away and moving away from me, but I saw was a hairy animal covered in longish, matted unclean-looking brown hair, approximately the color of an elk's coat. This animal was moving away at a high rate of speed, it was approximately 8 to 8 1/2 feet tall, approximately 3 to 3 1/2 feet wide shoulders. I followed it for approximately 80 yards with my spotlight up a ravine, uphill, it covered this 80 yards in approximately 4 to 5 seconds. The animal was not moving in a jerky motion, but had a very fluid stride and approximately 5 to 6 foot span between his strides.

Returned to the seen the next day, did not find any tracks due to the rain, however I did find where it went through a patch of buck brush and did break the brush, no hair or other physical signs at the scene. I came to my realization as to the height from the comparison on the tree that it went by, where the animal's head was approximately 2 feet higher than mine.

As I said, I followed this animal for approximately 80 yards with my spotlight, it went up the ravine, uphill, covered this 80 yards in 4 to 5 seconds, moving at an extremely high rate of speed, going up the hill, it was running in a straddly type motion, throwing its legs to the sides, very rarely bending them. It had long arms, it was extremely powerfully built. The feeling I got from this animal was that it was extremely powerful, possibly dangerous. I did not exit my patrol vehicle at the time.

I believe that what I saw is what is called a Big Foot or Sasquatch.

Dillon's encounter is incredibly fleeting— what, up to 7 seconds, 10 seconds? Not much. Although armed, he instinctively knew to stay in his car. He describes a hip centric type of bipedal walk, that rapid outward kick and throw of the legs instead of a smooth bending of the knee like a human. Most of the movement is therefore from the hip— quite a contrast to the Patterson Film, but it remains a compliment to the work of sober anatomists.

John Green, who had been mailed the affidavit by a co-researcher in Bigfootery, reproduced Dillon's statement in his self-published updated *On the Track of the Sasquatch*, Book 1, in 1980. He added that "Because it is an official account of close, clear observations by a professional investigator, including action that would be beyond the physical capacity of a human, it ranks as one of the very best of all reports."

Unfortunately, it is also the last of the very best reports. And this, of course, gives you an idea of what the reports have been like since 1980. Dillon was one of the last to make a clear, good report because as a sheriff he was patrolling at night, the time in which the Sasquatch is on the prowl. Not too many average guys are doing that at night, especially up in the mountains. There is therefore also little chance that credible observers would be in a position to see a Sasquatch. As the 1980s progressed the chances got smaller. That decade was not the "get back to nature" decade. Quite the opposite. It was the decade for a flurry of carnival reports, psychic sex encounters, and UFO aliens walking side-by-side with Bigfoot, a carnival where an average and reliable person was even less likely to be disposed to mention their encounter.

Publicity ebbs and flows. Even when the story is true it can go into overkill and the media finds no reason in reporting "old news." When that happens the froth fizzles away too. For Bigfoot there is left only an empty glass, no brew of any kind.

So empty is the glass in the last 30 years that John Green, the preeminent chronicler of Bigfoot, in his recent *The Best of Sasquatch/Bigfoot* (2004) could devote only the first short chapter to "Recent Developments," which surrounded Dr. Jeffrey Meldrum's research at Skoocum Meadows near Mt. St. Helens where he fancied any number of Bigfoot butt marks in the ground (euphemistically now known as Buttfoot). The second chapter was devoted to countering the claim that Ray Wallace hoaxed the Jerry Crew tracks in 1958. The rest of the book is simply the reproduced 1980 reprint edition *On the Track of the Sasquatch*. There is in fact no real genuine and credible Sasquatch sighting report since Dillon's brief encounter in 1979.

Mercifully, Green admits that most of the many reports being sent to research organizations on the web today "are obviously from pranksters, and trying to sort out the less-obvious fakes from the genuine information is a major task." Unfortunately, he then admits that "the most interesting thing about the flood of new information, however, is that the majority of the reports do not come from the traditional areas at all. There are far more reports from east of the Mississippi than there are from the west of the continental divide." This alone should have alerted him that the information is now, as it was before, mostly from pranksters. The only difference is that in the Pacific Northwest Bigfoot is old and suffering from overkill. A prankster's report won't even get a rise there anymore.

There isn't much to commend Bigfoot's presence in the other states, nor the method by which it is reported anymore; and there wasn't much back in the Pacific Northwest 50 years ago during all the hype. These recent developments are so sadly in-keeping with the Bigfoot copycat pattern that they require that we look at just what were the qualities of those who searched for Bigfoot during his heyday. They accepted and followed a sad pattern of copycat sightings, perpetuating in consequence the ludicrous and the inaccurate. Bigfoot was a dozen different looking creatures that had enormously different enormous feet. No Bigfooter tried to analyze it all and sort it out. And if they did, they used as their rule the foot that suckered Jerry Crew and those that were left by Patterson's Bluff Creek Bigfoot.

There was one occasion in which a hoaxer added even more brilliantly to the Bigfoot lore, and in doing so showed how easy it was to lead the Bigfooters in his tracks. And to this event we must go, this momentous— indeed, pivotal— event in the history of Bigfootery.

To come back to America is indeed to come back to the carnival. We must leave behind the Russian information, the accurate reports, the Canadian prints, and put on boots to follow a clown in enlarged human feet and the men who used Bigfoot for their own venal ends.

CHAPTER 12

# Anatomy of a Debacle

There is no better way to highlight the flummery of Bigfootery than by probing into the very event that was fundamental to the creation of its monster, *Gigantopithecus*. The Bossburg Incident is the fulcrum of the Janus, the Roman god of beginnings and doorways. He was portrayed in Roman art as two heads joined back to back, each face gazing in the opposite direction. The Janus Point, that connective seam between them is where the transition is made from old and new, that brief moment in the doorway frame where one exits the past and treads nigh unto the future— hence the symbol of two heads and doorways. It is at Bossburg where all that was original and old and genuine Bigfootery died and where a new media-based Bigfootery emerged.

The Bossburg fiasco was an outlandish free-for-all hunt like an Irish Boxing Day brawl. The object of the brouhaha was the much ballyhooed "Cripple Foot," the name given to a dubious "Sasquatch" footprint and the phantasm that must have owned it. There are no two greater pieces of evidence, let's say pillars, in the not-so-hallowed halls of Bigfootery than the Patterson Film and the Cripple Foot cast. They are indispensable as evidence and used more than anything to prove the Gentle Giant Bigfoot of modern lore.

Although the Cripple Foot cast is much talked about, and was even displayed for some time in the San Francisco International Airport's Bigfoot display (thanks to Grover Krantz), the circumstances of its discovery have been reported accurately only once. This was in Don Hunter and René Dahinden's 1973 book *Sasquatch*, the only book that was ever written from Dahinden's rather cynical and entertainingly sarcastic vantage point on the topic of Sasquatch.

The Bossburg Incident began in November of 1969 with a phone call from John Green to Dahinden, who was at the Big Horn Dam site in Canada where he had been investigating a number of reports of giant hairy Sasquatches. Green relayed information that at Bossburg in Washington State there had been some interesting developments. Ivan Marx had called him, Green said, and told him that he was following tracks made by what appeared to be a crippled Sasquatch. Both Dahinden and Green knew Marx from the old Tom Slick PNE days. Marx was not one of the original founders, but he came in later. As a professional woodsman and guide he knew how to make his way around the forests, so it wasn't long before he got involved.

Green was a bit monotone over the whole ordeal. Dahinden, too, by this time shared the same lack of enthusiasm that all old Bigfooters had over a simple track report. But Dahinden decided to call Marx to confirm. Marx's excitement was palpable. He reported that the tracks were first seen by the local butcher of nearby Colville, Joe Rhodes, on November 24, in soft soil at the community garbage dump. Since one foot—the right one— appeared severally crippled, Marx suggested that this Bigfoot was forced to live off "kitchen scraps" instead of foraging for its own food. The implications were obvious. This Sasquatch might indeed be trackable and catchable.

There wasn't one of the old Bigfooters who thought Bigfoot would be an easy target to fell anymore. In the beginning they had thought they could follow a trail and "grab Bigfoot in the ass" where it ended. By now it was plain as paint that it would either be by accident or by finding a corpse. "Cripple Foot" might provide an acceptable compromise. In the wild outdoors it's always the straggler or the wounded that gets picked off the easiest.

Dahinden considered the evidence for a few days and then went to Bossburg. The tracks were terribly trampled by all the locals who had come in the interim to look. However, someone had covered one of the right footprints with a cardboard box and in doing so had preserved it. Dahinden photographed it and then made a plaster cast— the true "Cripple Foot" cast from which the copies today are taken. Undertaking a local investigation to get a feel for what had been going on, Dahinden discovered there had been a preamble of sightings, though they remained vague and dubious and had also been dismissed by the local Sheriff.

Bob Titmus now suddenly arrived to take a look at the dump site and at Dahinden's plaster cast. It was plain enough from what he did next that he had had a long talk with Ivan Marx. He went right out and, according to Dahinden, "bought an eight-pound slab of beef and hung it in a tree." Titmus, naturally, was certain that as a crippled Sasquatch it would have to live off the dump. Titmus stayed out there in his paneled truck during the nights and watched it, waiting for the crippled Sasquatch to come along, "thinking that if it *was* a cripple and *was* living off the garbage dump, when it came along he would just grab it in the arse and throw it in the truck and run on home with it."

An amateur hunter named Norm Davis now appeared. He was a radio station owner turned newbie Sasquatch hunter. He now hung a basket of fruit in the neighboring tree. Davis was careful to measure off 6 feet and suspend it at this height so that Sasquatch would surely see it, and he would also know that no other critter could disturb it except the hairy giant.

The days passed and the meat was now crawling. In a huff, Titmus packed up and went back to his home, a 700 mile drive back to Kitimat, British Columbia. The business side of Dahinden made it possible for him to stay on at Bossburg in Davis' trailer. He cut a deal for his rent. He would show the Patterson Film, which he had rights to (having paid his half of $1,500.00 smackerals), and would talk in local service clubs. Presumably, Davis would get the proceeds. They moved the trailer to Ivan Marx's property and began to formulate plans to lay in wait and get the crippled Sasquatch.

On December 13 the most surprising thing happened. Dahinden, Marx and a local man, Jim Hopkins, drove along the banks of Roosevelt Lake, the reservoir of the Grand Coolie Dam. Dahinden had placed out scraps of fresh meat here and there some days before, and everyday he checked to see if Sasquatch had been baited to the scene. Today, as they cruised along the three kept their eyes peeled for tracks in the soft earth. Yet, per usual, nothing was sighted except, unfortunately, a parked jeep, which meant somebody was out there right now walking around. That always made Dahinden suspicious, since by now he was so jaded by hoaxers.

After continuing for 4 more miles they were now close to a railway crossing where the road and the railways ran close to the Columbia River. Ivan Marx decided at this point to hop out and look around while the other two opted to stay inside. Amazingly, "Marx was away from the car only seconds before he came racing back" and shouted "Bigfoot tracks!"

Dahinden did not respond, but according to his book with Don Hunter he remained filling his pipe and looking over it at Marx. "Okay, joke over."

Even more amazing, instead of immediately urging them out to investigate, Marx, a feverish amateur photographer, jumped in and whipped the car around, saying that he needed photographic evidence. As Marx was rushing them back home to get his camera, Dahinden was far from excited since he hadn't even been given time to get out and look. Already fearing the prints were hoaxed, he cautioned them to scan inside the parked jeep as they passed to see who was in there. He also told Jim Hopkins to get the license plate number.

After Marx secured his camera equipment, and René his rifle, they were back at the scene. Sure enough, almost on cue, there were the Bigfoot tracks. It was Cripple Foot! The right foot measured 17 ½ inches long and was 6 ½ inches across the ball, tapering to 5 ½ inches across the heel. Dahinden was impressed, and cocked his gun. They followed the tracks and carefully counted their number. There were 1,089 definable prints.

At this moment Dahinden instinctively regarded the Cripple Foot as the best evidence for the existence of Sasquatch. But there were some things that were to happen that would make him more cautious about the whole incident even before the carnival that would become the Bossburg Fiasco was to break loose.

From Hunter and Dahinden's account of it, I can see why. The tracks led through the mud and snow from the river, over the railroad and across the road, over a fence (43 inches tall) into more snow and then under a tree on ground free of snow. Here there were no clear tracks. There were only depressions made by the feet and nothing more definable. But "right in the center of a depression [of tracks] was a clump of snow holding the imprint of the toes of the left foot, as though the snow had been shaken loose after building up on the foot."

This should have alerted any hunter that this could not have been an animated foot, since the mobility of the toes would not allow for a big clump of snow to remain in place after walking for a distance on the mud and pine needles under a tree. A fake foot, on the other hand, could hold a build-up of snow around its stiff toes and easily carry this for several feet until falling off in a clump. We must remember that all around there had been a fresh and therefore soft snowfall. It would be easy for a normal sized man to make deep depressions in the snow; but the prints, as recorded by Hunter and Dahinden, in the dirt under the tree were negligible and amorphous and only referred to as "depressions."

Regardless, Dahinden was sure they were now "going to get that hairy son of a bitch."

In following the tracks up the snow covered hill something truly suspicious was then discovered that sullied Dahinden's enthusiasm. They thought they would catch their quarry soon. However, the tracks led only about halfway up the hillock and then turned and retraced their ascending path, anywhere from about 40 to 50 feet apart from it. At one point the pair of feet were side by side indicating the creature had stopped and stood. Between the pair of footprints they discovered a "deep yellow patch" as it apparently had stood there and wizzed up a storm— something that can happen to any of us after an unaccustomed and exciting workout! None of them decided to take a sample, although they thought it was urine.

They continued to follow the varlet's tracks back down. They crossed the fence 50 feet from the ascending tracks and continued on to the railroad. Then they came back to the fence and over it and back to the railroad as if the creature was zigzagging back and forth. Then the tracks led back through a patch of bush before they then stopped at the steep part of the river bank that overlooked the river from a height of 150 feet. The tracks then turned upstream along the bank, where they continued for about 200 feet, going down the gradually sloping bank to the river's brink. There they stopped. There was evidence that one foot had slipped on the slope while going down to the river's edge. This was determined by a deep groove, as if made by a slipping heel.

That was the Bossburg Cripple.

No more again was it seen. The tracks conveniently appeared where Dahinden had been consistently checking on a daily basis. He had been looking just the night before and noticed there was nothing. He knew therefore that these were fresh tracks. This also made Dahinden skeptical.

The very fact that the "creature" simply walked up a hill and back *should be* suspicious. It more or less looked like a hoaxer was laying a trail. Moreover, to intentionally walk back and forth to the fence and jump over it, to make it look like the crippled Sasquatch was stepping over it, seems to be overkill, trying to make as many point-less tracks as possible, to impress upon the amateur that it was *not* a pointless walk up the hill and back. That isolated clump of snow with toe prints should have been regarded as one of the most suspi-cions clues of them all.

Others came forward within the next few days and claimed to have seen the tracks. A Border Patrol officer said that he had seen them on the other side of the river in land owned by timber compa-nies. Due to the fact the access gates to this property were usually locked some thought this amplified the mystery, for why would a hoaxer go and leave tracks where no one would see them? That is a statement of illogic. The reason, of course, could be that our varlet worked there and wanted to take a practice run before going to pub-lic land.

Later while in Bossburg, Dahinden and Marx were also able to identify the men in the parked jeep. They, that is, Dahinden and Marx, asked them if they had seen the tracks. Their "yes" was defi-nite, and then they added emphatically that they immediately "got the hell out of there." This should have proved even more suspicious to Dahinden, Hopkins and Marx, but none of it did. The three did not take into account a very salient fact. None of the three fearless Bigfooters make the statement that they ever saw any other man's footprints the entire time they followed the "Bigfoot" tracks. From where then did those two men see the tracks close enough to identify them as Sasquatch and hence to be so intimidated?

René Dahinden should have clung more to his original suspicion regarding the Bossburg Cripple's circumstances and used this as a mitigating factor in his forthcoming decisions. But he seemed to catch the growing fever that they might indeed be on to a Sas-quatch, once again merely because the "crippled" footprint was so different from the perfect footprints that had hitherto been found at Bluff Creek.

Another mistake was that he did not cross the river to try and pick up the trail to see where it would lead. Instead he retraced his steps and followed the footprints back over their course up the hill, doing this 7 times, examining the details of that "crippled" foot to a mes-meric degree. It was this feature alone that would hypnotize so many.

According to Dahinden, the crippled foot showed greater com-pression in the snow than the left foot. He unfortunately deduced the opposite of what this would logically imply. He thought it meant

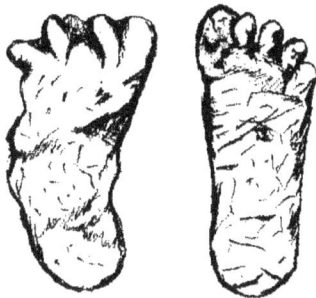

*The notorious Cripple Foot. The left foot is merely an enlarged and very flat human foot.*

the creature was favoring the crippled side; that this leg was shorter and therefore more weight was being placed on it. If so, the crippled creature was favoring the wrong side. The left foot should have been deeper since it was not the crippled foot and any truly crippled creature would not place the greatest weight on the injured member.

Dahinden and Marx lifted some of the prints from the snow and stored them in Marx's freezer, before which Dahinden was able to vegetate and examine them.

This was hardly the end of the Bossburg incident. "Dickie" (Norm) Davis was disturbed by how Dahinden merely brooded over the prints, thinking and thinking about every nook and cranny in them (this was Dahinden's style). Davis was eager to go find more tracks. He called Titmus and other Bigfooters and told them the news. Titmus was back, again driving 700 miles to return to the promised Bigfoot land. These phone calls also brought Roy Fardell from British Columbia and a young zoologist from San Francisco, Roger St. Hillaire. They too brooded over the tracks while Dickie Davis fidgeted about.

There was a disastrous consensus: since the tracks continued on to the other side of the river (which Dahinden had confirmed on site days later) the crippled foot tracks had to be real. The illogic of the hypothesis of why would a hoaxer continue to go into an area where the access roads were closed and where there was little chance to find his tracks prevailed . . . sadly.

Others were converging on Bossburg. Pretty soon the area was trampled by tourists, curious locals and by a number of Elmer Fudd hunters.

None, however, were as venal as the old Bigfooters. Now Roger Patterson and an associate of his, Dennis Jensen, arrived. Patterson would come and go over the drama's unfolding, but he left Jensen there to protect his interests. This would amount to quite a period of time, for the saga continued on into January 1970. For these two months hunters would be plowing through the dense forests with snowmobiles, plodding on snowshoes, galloping over it with horses and just plain staking out a place behind a tree and waiting to ambush the Sasquatch.

Even the cold blue skies boasted an airplane droning back and forth, as one of the main Bigfoot camps kept an alert eye on the others. That is, in due course, what most of the surveillance was channeled to: watching the other guy. One local hunter even complained in the local *Colville Examiner* that he couldn't even step behind a tree and "take a leak without feeling a dozen eyes watching me." Whoever got Bigfoot first was going to have to fight to keep him!

It is interesting to note that not once in these two hectic months did anyone from this army of hunters see any more Cripple Foot tracks. Someone clearly realized it would be dangerous to be out there.

Dahinden was now beginning to suspect they were following "a phony," for "the longer they went without finding a trace of the cripple, the worse the suspicion grew, and the more the others began to share the doubts." Instead of suspecting a hoax, a new suspicion went around the group of Bigfooters. Did somebody already have it bagged and was he hiding it? Like some mystery theatre drama, the distrustful coterie of Bigfooters imagined that the guilty party was hanging around until everybody got tired and left, otherwise he would be tipping his guilty hand.

Finally, by the end of January 1970 Dahinden decided to pack up and go try to peddle the Patterson Film on a Canadian lecture tour, heading up to Calgary as his first stop on the circuit. Dahinden vowed to stay in close touch with Fardell and St. Hillaire, who were squarely in his camp during this venture.

Bossburg was still clamoring with the old Bigfooters, who now were making Marx's home and Davis' trailer their headquarters.

Something now happened that was in-keeping with the general undercurrent of suspicion. Suddenly two men arrived at Bigfoot HQ. They were a prospector, Joe Metlow, and a state wildlife agent, Bill Streeter. Streeter was calm and a bit detached, as if merely accompanying his friend. Metlow was to the point. He asked how much would the body of a Sasquatch bring— to be explicit, how much are you willing to pay?

Patterson's partner, Jensen, with an air of seedy grandeur, nobly told the gentlemen that the value in money for a Sasquatch was not the issue and it didn't come close in importance to the need of scientifically establishing the reality of the species. Jensen then informed Metlow of the very real difficulties of actually getting one, reminding him how long many of them present had been trying. Apparently, Jensen had assumed that Metlow was another amateur who just arrived and wanted to pick their brains. Far from the fact. Metlow now interrupted him and said: "But I've got one, and it's sold."

Don Hunter writes with a good degree of humor on what happened next. He said that "nobility and science fled." Jensen immediately replied "We'll top your best offer."

Metlow now told his story. On January 27 while out running claim lines on a gold prospect (a place he kept secret) his path and that of a "cream-colored" Sasquatch met. The Bigfoot was digging about at ferns, and Metlow thought that the beast was building a home for itself. At this point Streeter came forward from the timid shadows and confirmed that Metlow had told him this as well, and then he withdrew from any other role in the conversation.

The gist of Metlow's long story was that he grabbed the Sasquatch by the gonads and threw it down a mineshaft where he knew he could keep it alive. He never got specific and never divulged who had given him a cash offer. It wasn't necessary to bait an eager and gullible Jensen anyway. Jensen quickly replied that the only money in the field right now was Northwest Research Association— i.e. Roger Patterson's front in Yakima, Washington.

Jensen made it all sound very grand indeed, saying he would phone the "principles" of this organization, which was, of course, Roger Patterson, the inventor and promoter of precarious living, to use John Green's words, who had produced the highly dubious Bluff Creek Bigfoot film. Patterson, of course, had no money. He was sucking on the pap of an Ohio merchant named Page, feeding the curious Ohioan reports for a yearly retainer. When Jensen informed Patterson of the story, Patterson immediately told Jensen to keep Dahinden's group out of the bargaining.

The produced an incredible feud! Fardell and St. Hillaire had snuck around like a couple of badgers and found out about a Sasquatch capture and a pending deal. They immediately called Dahinden in Calgary. When Dahinden got wind of this, the stink was on! He employed his Swiss accent to recite every 4 letter word in the book. He then told Fardell and St. Hillaire to continue to rummage about for information. This they did, discovering where Metlow had been prospecting on the Frisco Standard, a mountain view to the north.

Meanwhile Patterson had been electrified by Jensen's account. He had talked to his financial backer, Tom Page. Page gyrated and hopped on a helicopter, arriving in Bossburg to meet with Patterson (who had secretly returned from Yakima), Jensen and Metlow. Whatever was agreed upon, Patterson, Jensen and Page moved their headquarters to Colville, Washington, some twenty miles away from Bossburg, no doubt trying to stay clear of being spied on by Dahinden's allies, Fardell and St. Hillaire.

Despite some tentative agreement, Page, Patterson and Jensen were still ignorant of where Metlow had been prospecting and therefore were in the dark as to where he might have seconded his albino Bigfoot with the sore gonads. Page was now offering $35,000 dollars for the Sasquatch, but apparently this was not enough to tempt Metlow, who continued to remain coy.

But Page was sure he would win. He knew that Dahinden didn't have enough money to match him. So Page had a helicopter gassed up and on standby ready to go into action and fly up to the mineshaft to retrieve the Bigfoot at the moment Metlow gave in.

Fardell and St. Hillaire were calling Dahinden in Calgary daily to keep him informed. Dahinden ordered them to stall and keep the opposition covered. This they did quite religiously, staking out the Patterson camp and forever being ready to move at a moment's notice should that helicopter's blades start to turn.

A local newspaperman, Denny Striker, was also clandestinely involved. He was finding out about every move either camp was making. He was also staking out Patterson's motel, once for 72 hours straight. His jeep was always hidden in the brush ready to go and get a scoop when the Sasquatch was retrieved.

Both Norm Davis and Ivan Marx tried to stay neutral between the two camps despite the fact that in between stalking Jensen and Page, Fardell and St. Hillaire bunked at Davis' Bigfoot HQ on Marx's property. However neutral Marx was trying to remain, he was nevertheless overhearing the numbers that were being bandied about during the bidding war that ensued for Metlow's Bigfoot. Dahinden, after having talked over the phone with Metlow and, finding him cagey, had called St. Hillaire and Fardell at Marx's. He told them the prices were going up. Hillaire and Fardell were ordered to outbid Page.

There was terrible friction between Dahinden's camp and Patterson's camp, and this became grossly obvious to the whole community of Bossburg. Now it was coming close to February. The Bigfooters had originally been town celebrities during their stay, but now

[248]

the town was more or less disenchanted with them and regarded their presence like a wart on the town's fanny. Seeing different camps trailing each other and spying on each other was tawdry and disgusting.

When Patterson and Page finally got Metlow to uncork that it was Frisco Standard where he had been prospecting, Patterson and his camp raced up there in snowmobiles only to arrive, turn off their engines, and hear the sound of an airplane overhead. It was St. Hillaire and Fardell watching them, making it plain by wobbling their wings.

Yet neither camp found anything.

Amazingly, this didn't seem to make any of them remotely suspicious of Metlow.

Indeed, the bidding war escalated while Metlow remained coy. It was up to $50,000 when Dahinden, still packing up in Calgary, called John Green and asked him to get to Bossburg and protect his interests. When Green got there, however, even Jensen was surprised that Green spent most of his time, as Hunter phrased it, "galloping between the camps to see where the best deal was for him." There would be no more close Dahinden-Green partnership anymore. Still in Calgary "without a pot to piss in," Dahinden finally offered $55,000 dollars, the top offer, which he admitted was a total bluff. Hillaire and Fardell gave Metlow the offer on his behalf.

Metlow was still coy, and Page and Patterson were glad of it. They still had a chance to consider their next move. This allowed a certain respite between bidding, and in this interim Dahinden got to Bossburg. This led to a slight truce as both sides tried to figure out why Metlow was still coy. Patterson's and Dahinden's respective groups were both able to meet at Metlow's home and discuss the deal. Everything went well during very "genial" conversation until Metlow said he had a Bigfoot foot in his freezer. Dahinden immediately offered 500 bucks just for a look at hit, and Metlow soon haggled him up to 5,000 dollars.

I defer to Hunter's narrative: "The excitement brought Dickie Davis to the scene, antennae quivering. Before you could say 'Bigfoot,' Davis had a contract sketched out. It would include John Green to write the book, Bob Titmus to skin and dissect the foot's owner— presumably stashed in a cave somewhere above the snow line— and anthropologist Grover Krantz to introduce it to science. René was out, so was Patterson."

Metlow now confessed the foot was in his sister's freezer in Portland, Oregon. Like a flash Davis sprung for the tickets. Davis and

Metlow left for Spokane and then from there they would head by plane to Portland. But in almost movie-like staging Metlow slipped and hurt his ankle and had to stay behind. Davis went on his own, leaving Metlow at the airport. Metlow said he would get his ankle fixed and catch the next flight out.

Davis havered about Portland for two days before finally lighting the embers of his mind. He came back and scowled at Metlow's excuse that he had been "real sick, Dickie" and couldn't make it. In reality, he had immediately cashed in his tickets for what Hunter said was to buy "a couple of bottles of the best tasting pain killer" and went home and watched TV.

It was now obvious that Metlow was a crank. All the Bigfooters packed up and left Bossburg. The Press had followed enough of the shenanigans to give the whole pastime of Bigfootery a deserved black eye. It was obvious that no one was really interested in getting a Sasquatch for any scientific or noble reasons. It was prestige and money.

Jack W. Ondrack, in a paper presented at the Western Association of Sociology and Anthropology in Banff, Alberta, Canada, had this to say in December 1969 as all this was playing itself out before the press:

In the recent past, Sasquatch research has been conducted by poorly financed, untrained, dedicated men. A Sasquatch hunt with these people reminds one of a Humphrey Bogart movie in which a number of individuals having idiosyncratic and conflicting goals cooperate sporadically to bring a mutual goal closer to attainment. This goal however is mutual only in the sense that each man wants to find a Sasquatch, and not in the sense that each man wants somebody to find a Sasquatch.

Thus ended America and Canada's fascination with Bigfoot hunters and thus marked the beginning fascination with the creature itself.

But if Cripple Foot was completely lost and forgotten in all this, another stunt was about to resurrect him and bring some hunters tramping back to Bossburg.

# CHAPTER 13

# Cine du Sasquatch

The maxim of the Comedie Française is true: "when the audience laughs with you, you are a success; when they laugh at you, you are ruined." The Cinema of Sasquatch is a montage of comedy, drama, suspense, and always farce. Thus it is a perfect reflection of Bigfootery. It is a world where faux pas is studied as science. It began with Patterson and reached its most ludicrous with Ivan Marx. The ability of everyone to see through the farce at Bossburg except the old Bigfooters is what destroyed Bigfootery's credibility in a hail of laughter. When new film would now be produced of the "Cripple" by Marx the last bit of veneer Bigfooters had was gone forever. Their gullibility reiterates with stunning clarity and resounding authority *why* a true mug shot of the "animal humans" was never drawn and why therefore a Sasquatch has never been found.

Inconceivably, Dahinden was still interested in the Bossburg "pulse" of things as it related to Cripple Foot, and kept in close touch with Ivan Marx. Marx had apparently admirably played no sides in the debacle, seeing that there was nothing to be gained. However, it now seems he might have been fomenting other plans more closely aligned with Cripple Foot's debut on camera.

Marx was now finding loads of evidence for Cripple Foot, enough to keep Dahinden piqued. There was a handprint found here, a footprint there or evidence where Sasquatch bedded down, or whatever. Yet Bossburg remained off the map. No one was coming back.

Then in October 1970, almost a year after Cripple Foot had debuted and never returned to the stage to take its bow, Marx called Dahinden. "I've got a film of the Cripple." This was not the only person he contacted. Denny Striker was also notified and all the details were in the Colville *Statesman Examiner.* Before it hit the Press, Dahinden was in Bossburg and the other Bigfooters were preparing the Charge of the Fright Brigade back to the woods.

The article, far too coached and gullible (if not outright in collusion) read that Ivan Marx got a telephone call from an "unidentified person" claiming "a large creature" had been hit "by either a car or a train" about 7 miles north of Bossburg. When Marx received the call he immediately packed up his dog and camera with the hopes of actually following the creature. The article stressed that Marx went unarmed except for his Bolex 16mm movie camera and a 35 millimeter Nokon camera and, lastly, a two-way radio so he could stay in touch with a local rancher, Don Byington.

The weather was heavy overcast and smoky when Marx came upon the creature and "jumped" it at the bottom of a dense draw and immediately began filming. "The initial footage shows a large black upright figure moving stealthily but rapidly through the dense growth, but only in silhouette. . . .Marx pressed the pursuit with his hound, forcing the Sasquatch into a clearing where, with his movie camera set at f2.8 he took the remarkably clear footage of an impressive looking creature."

On the screen the Sasquatch is shown moving from right to left at an angle of about 45° away from the photographer. Distance from the subject according to Marx ranged from 25 feet to more than 100 feet as it made its way into the heavy underbrush on the far side of the clearing.

Probably the most impressive part of the film, besides its extreme clarity, is the fact that the Sasquatch is visibly injured, holding its right arm tightly to its chest and using its long muscular left arm for compensating balance.

Also, both ankles of the creature seem badly skinned, the wounds showing plainly raw against the black hair of the legs and feet.

In watching the frames singly, the injured or skinned area appears to extend onto the bottom of one foot, and possibly on both feet, which would account for the apparent pain-filled moments of the frightened creature.

As the Sasquatch is nearing the far side of the clearing, a twisted tree limb is stepped on, bouncing up and striking it above knee level. Marx, the following day, photographed this stick which was 10 feet long. In comparison the creature photographed would have stood about 9 feet tall and Marx estimated its weight as that of two large bears, or around seven to eight hundred pounds.

The only thing the film is lacking is facial features on the creature. Twice while crossing the clearing the Sasquatch turned its head to glare at Marx. The first time it turned 180 degrees and uttered a weird scream which was heard by Byington, positioned on a ridge nearby. The second time it turned a full 360 degrees, appearing quite confused, but the lack of light prevented any facial features from showing plainly.

Marx said he continued pursuit of the creature until darkness prevented further advance, and when the trail was recovered the following day, it led through a maze of rugged terrain and finally to a body of water where it was lost. He feels the Sasquatch is very old and apparently hurt quite badly.

Since Dahinden had rushed and was already at Bossburg, the article contained a quote from him declaring that "Ivan has a movie and that leaves only two choices. Either it's real or it's not. That's what I'm here to find out."

This time there was no dickering around except by Dickie Davis who quickly arrived and, without even seeing the film, offered Marx money. Ivan Sanderson was also on the line for the men's journal *Argosy*, enquiring about still-picture rights. Page was quickly there and offered Marx $25,000 for a copy, sight unseen.

The film is so pathetic that it is remarkable that most of the Bigfooters who converged on Marx said it was the real thing after being allowed to view it. Even John Green quickly declared it authentic upon arrival at Bossburg, leading him to write a "tribute" to Marx in the *Bigfoot Bulletin*, declaring that Marx could not have "faked all he has to show, and that the film is genuine." This afforded Dahinden some amusing diversion. He subtly criticized Green's gullible outlook and inept and shallow methods of investigation, a pastime of criticism he would continue the rest of his life.

The problems were not just with the ludicrous film. The same problems with Metlow were now being encountered with Marx. He refused to divulge the location where he shot the footage and he remained coy about accepting any money. Had any of the Bigfooters bothered to pay attention to Don Byington's two kids they would have found a gem. They had been murmuring something very interesting: *they knew* where it was filmed.

No one listened to them until Peter Byrne arrived at Bossburg, ready to investigate the claims. No one had seen the film commercially because Marx wouldn't show it except at his house. Byrne now offered an agreement which required Marx to use the film as security, but he did not have to commit himself to rights over it yet. Marx was to put the film into a safety deposit vault with Byrne's Washington D.C. lawyers, remaining unexamined (though Byrne had seen a copy of it), and he could then draw a monthly retainer of $750 while he considered the prospects of who to agree to sell rights to. Marx was happy, and this arrangement went on through March 1971. Marx continued to be subsidized to hunt Cripple Foot and he had to make no commitments to whom to sell the film rights.

Unfortunately, Byrne came across Byington's two children. They still insisted they knew where the film had been shot. Byrne went with them to the back of their own property and indeed Byrne found the same location as in the film. There was the identical tree the Sasquatch had passed under and the branch where its head had scraped as it did so. Marx had said that the branch indicated Sasquatch was 9 feet tall. Byrne was flabbergasted. The branch was less than 6 feet from the ground! He examined the area more closely and realized that Marx could not have been shooting the film from the angle that he had claimed. Nor did he use a regular lens as he had insisted. Byrne could prove he used a telephoto and that what Marx filmed was a long distance shot of somebody less than 6 foot tall. Byrne then found out that Marx had been buying large pieces of fur at intervals in Spokane.

Byrne and his associates got together. "We had a brief meeting," recalled Byrne, "and then decided that it was time to have a serious talk with Mr. Marx, to ask him, among other things, how he could have been mistaken— to put it kindly— about the site of his filming . . .Alas, we were too late. In the night— as was clearly indicated by the discarded personal belongings strewn across his front yard, our quarry had got wind of our plans and, as they say, had upped and run for cover. And in a hurry, too, leaving a veritable river of trash running from the open, flapping-in-the-wind front door of the shack to where he parked his VW bug, one that included ancient and tattered magazines and newspapers, old patched and repatched gum boots, torn cotton towels, plastic rain coats, ragged shirts, woolen hats, ripped up, oil and grease stained work shirts and trousers, empty motor oil containers, rusting baked bean and soup cans, stained and ragged blankets, mayonnaise and jam and pickle jars and dog food cartons and half filled trash bags." When he found Marx had fled Bossburg, Byrne called the lawyers to check his security— the film. It was nothing but black and white footage of old Mickey Mouse cartoons. "Not even color at that!"

Marx, like Titmus, Dahinden and Green, had been one of the old Slick Expedition's (PNE) members. Now that the 1970s were breaking, Bigfoot was becoming a world famous character. But for the old hunters it had been 13 years and still nothing had been found. The temptations of money and publicity were now obvious, and it had been their desire all along. Patterson had raised the stakes by his famous film in October 1967. But the others had not been able to ante up. Even Dahinden was brutally honest about himself in *Sasquatch*. He claimed he was shattered when Patterson took the 1967 film because, as Hunter writes, a "virtual novice would collect the glory and the hard cash prizes that he had always assumed would accrue to the first finder of Sasquatch, and whom he had always determined would be himself."

I don't know if these old Bigfooters had watched too much of *King Kong* or what else motivated them to think that they were going to catch a Bigfoot and be allowed to put him on display in a carnival or at the Palladium or Carnegie Hall and reap ticket money. There would really be no profit in catching Bigfoot. But their showman mentality reveals a rather naïve frame of mind and a gullible presumption that is also sadly reflected in their entire investigative process on and approach to Sasquatch. Only such a mentality could accrue the kind of worthless data which they constantly gleaned, acquired by a gullibility which more than anything created the modern myth of Bigfoot.

How many tracks had some of the old hunters faked or at the very least been easily duped by? None of them displayed keen investigative talent, although Dahinden's natural contrariness helped him to ferret out some real cranks. Before Patterson's film, no big money had been offered. Faking tracks was good only for maintaining interest and keeping oilman Tom Slick involved. (Dahinden said that PNE members did engage in hoaxing). When he died in 1962, so did the tracks (except for Wallace') and most of the old PNE member's interest. None of that revived until 1967 and Patterson's film, and that was the crest of the wave that was still carrying them past Bossburg. Frankly, with the help of Cripple Foot that carried them through the 1970s and into the 21st century.

Despite Dahinden's frequent caustic denouncements of Bob Titmus' ineptness and John Green's shallowness, he deserves little praise for his Bossburg investigation. Like Titmus at Bluff Creek, Dahinden never measured the stride of the "Cripple." There is no mention of it. Factual data, even by its conspicuous absence, is easily glossed over because there was something tangible to detour people's (and the Press') eyes— that Cripple cast. Patterson's film was enough for people to forget the lack of on-site evidence. The cast was enough at Bossburg. Even a wounded Bigfoot should make a stride longer than a man's. Wasn't anybody curious?

There was a minor epilogue to Bossburg which demonstrates Bigfooters' sad mentality poignantly. On Sunday the 21st of October Marx appeared on *You Asked For It*, the nationally syndicated show that probed into the odd and unusual. There he was for all to see clutching his treasured can of film of the Bigfoot he had just filmed in a snow storm in northern California. This was another white Sasquatch. After tracking it through the storm, he decided to head it off at the pass, like his other triumph, and film the Bigfoot as it came into view before him and his ever-ready tripod camera. It was then announced that this film had never been processed. Ivan Marx had just brought the film with him from the camera, and now for the TV audience the film would be processed by the producer and it would be shown to the world here for the first time on TV!

The audience was shown the tape of this process, and then the moment came. The film was clear and crisp and showed someone running in snow in a fur coat. "Great folds of the suit swung around like an old army blanket amid the California snows," wrote Hunter. It "cavorted" before the camera, ran toward it and then ran off through the snow, "flipping its clumsy feet backward and sideways as one does when running through deep snow in an overcoat." A primate specialist that the TV show had standing by rendered the official conclusion: "I think it's a man in a beast's suit."

Marx's hoaxery nonetheless proliferated, and such an emetic as this did not curtail his audacity. He would go on to make the ghastly high flop of a documentary *The Legend of Bigfoot* (1975) which has become a cult docudrama from the 1970s, and can still be secured from an internet distributor. This documentary is a remarkable example of Marx's self-deception. The Bossburg footage is in there

Most reviewers openly assumed it was his wife simply parading around in a bad monkey suit. Over this is Marx's serious voice-over narration attempting to huckster the audience into viewing this as extraordinary film. A road kill cat looks more like a Sasquatch, and even the kindest reviews of Marx's movie referred to him as a "cheapjack."

The Cine du Sasquatch had no greater and more transparent huckster than Marx. His pathos throughout the narration to make himself look like a qualified specialist and tracker, game warden, and government specialist in tracking renegades, one of a "dying breed" of men on the track to prove Sasquatch, is hilariously offensive in its overkill. Byrne, the only Bigfooter to financially retain Marx since Tom Slick, viewed him only as a "mediocre woodsman."

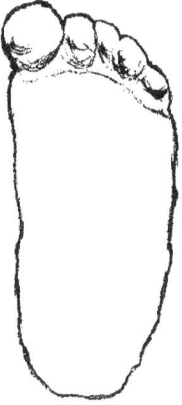

But the fact that Marx could survive as long as he did amongst the Bigfooters speaks of something else, something not so flattering to them. If Marx thought his outlandish 16 millimeter films could con movie audiences, it was because he had been lulled into this false sense of assurance by the low quality of investigation displayed by the other Bigfooters for so long. No one could look at his film for more than a few seconds without bursting out laughing at a man (his wife possibly) gimping in a monkey suit. . . all except the audience of old Bigfooters who studied the film; some of them, like John Green, even instantaneously declaring it authentic.

Marx Bigfoot foot

Even the internet distributor of such old lost films as this (Bijouflix.com) cannot help but comment in its advertisement that the Bigfoot believer should also "be prepared to feel like you were fleeced" by Marx.

Marx's later documentaries would showcase film from his many encounters with Bigfoot. One even showed Bigfoot waving at the camera and swimming through a lake. Audiences collectively laughed at Marx's ineptitude. If anything could inadvertently bring humor to Bigfoot's venal pursuit, Marx did by trying to seriously project his many faceless, bowlegged Bigfeet as real.

But there should be no question that Bossburg ushered in the new era of Cine du Sasquatch. It was such a media event that Bigfoot was on the nation's mind. Ron Olson's North American Wildlife Research Association scored a brief window of remembrance by taking

[ 257 ]

advantage of the new hype and producing (and being profiled in) its own documentary, 1971's *Bigfoot: Man or Beast?*. It was a 30 minute charming documentary in the mold of old school 1950s' hard-boiled one-reelers and 1960s' Disney wildlife escapades.

But in early 1973 a surprise low budget film *The Legend of Boggy Creek* came out by Chuck Pierce. It covered the alleged existence of a Bigfoot creature around Fouke, Arkansas, near his native Texarkana. The film and the Fouke Monster became surprisingly famous. But more important than this Pierce's goal to get the film in the theatres caused him to conceive of a new way to make a documentary. He used original "eyewitnesses" or participants and "actors" to reenact supposedly real events. This gave him enough film time for a theatre release and this gave us a new art form— "docudrama." *Boggy Creek* grossed over 22 million dollars, making this shoestring documentary not only the 7th most profitable film of the year but making it the father of docudramas, the staple of the 1970s' "reality" TV and film.

Following this format, Ron Olson earned panned reviews for the 1975 docudrama *Bigfoot: the Legend of Sasquatch* (incomparable nature shots, however!). Robert W. Morgan also entered the scene. He was the original inspiration for the TV series *In Search of. . .* which had been taken over by the most well-known documentarist of the 1970s, Alan Landsburg. Morgan's documentary *In Search of Bigfoot* carried his own recollections of his encounter with the Sasquatch. He reported that it looked like the "most man-like looking gorilla I ever saw." His 1975 flick is perhaps the archetypical docudrama of the 1970s' attitude about nature and Bigfoot's exemplary place in it. "Good God, we can learn from him!"

This was becoming the dominant attitude, the most pungent aspect of the S&M of "Bigfoot." And so by the time that giant of 1970s' off-beat documentary distributors Sun Schick Classic released a David Wolper docudrama *Mysterious Monsters* with Peter Graves as the host the berry-eating buddy qualities of Bigfoot were broadcast from coast to coast. Sun Classic's *Mysterious Monsters* (1975) was an incredibly influential documentary. To engage in a little harmless recollection here, I recorded mine on VHS in the mid 1980s when TBS would have such venerable weeklong salutes as "Mysteries Week on TBS" and would present one of these each night on the primetime movie. This is no longer done, unfortunately, and these old documentaries are almost impossible to find. Today, as I understand it, they are pirated out of China and do well on eBay on DVD. *Mysterious Monsters* is one of the most sought after and actually has an American distributor.

The point of this tangent is, however, quite relevant. It all began with Bossburg and then, via a strange detour, the "Fouke Monster" and Pierce's innovation made "Bigfoot" the pioneer and subject of a new and profitable style of film that was copied throughout the 1970s and into the '80s by any B-movie wannabe wanting to cash-in as Pierce had. Docudrama made Bigfoot the golden boy of the period. But it was the carnival creation of modern lore that walked over the silver screen. The golden boy walked far enough to lose the old violent image and acquire an entirely new persona complimentary to the 1970s' peacenik attitude. From the violent eerie human monster of the deep woods, he became our pure, primitive self, the archetype for the new age. Cine du Sasquatch had made it possible only because of the hype of Bossburg.

Who did make the Bossburg prints that started all this? Was it Marx to begin with, or did he merely capitalize on them? They never showed up at Bossburg again, nor likewise did any lead on them appear after Marx fled. Could he have been in cahoots? In unexplainable zeal to get his camera he fled the banks of Roosevelt Lake before Dahinden could even get out of the car. Did he know the men who were in that jeep? And, more importantly, did he know what they were presently doing out there? Was he giving them time to finish?

I would suggest that Marx capitalized on them. In a September 2003 article written by Peter Byrne, he recalled that while Marx was retained to look for tracks he reported finding a number of them "but for various reasons was never able to lead us back to them." Marx doesn't sound clever enough to have made the Bossburg print. But what about those two men in that jeep? They claimed they knew it was Bigfoot and got the "hell out of there." Yet they left no footprints in the snow to indicate they were ever close enough to the tracks to have identified them as Bigfoot. . .unless one was wearing the fake feet to begin with.

Dr. John Napier would suggest that the Bossburg prints have a better claim to authenticity than Patterson's Bluff Creek prints, but he was still uncertain. It was impossible that two different species of Bigfoot existed with human feet, and he could readily see that the Bossburg print was merely an enlarged human print whereas he had dubbed Patterson's Bigfoot print from its shape to be an "hourglass" type print (thanks to Wallace's fake foot). Napier apparently didn't know the circumstances of the Bossburg trail, a mere gallivant up the hill to take a wiz and then a back-and-forth zigzag over the road until it went back to the river.

From the circumstances above, I think it is clear that the Bossburg Cripple was little more than a bad hoax. Napier, though cautious, speculated that the crippled footprint impression could have been from an error in the casting medium used to create a fake foot. But far more certain than this, all the circumstances indicate a hoax. Those Bigfooters who wanted to accept it as genuine were perhaps motivated by the publicity that was being accorded Bigfoot in this new decade of the 1970s, post the 1967 Cine du Sasquatch genre started by Patterson. None seemed disciplined enough to really care about the circumstances. A new era had dawned, one of careless and often thin veneer, one where only the lightest things could float at the top. So light and effervescent was this new decade that a giant was born amidst the froth.

CHAPTER 14

## Jury-Rigged Giant

The inclusion of Grover Krantz in Dickie Davis' contract shows what a quick impact this physical anthropologist had on the Bigfooters. Just by his short visit to Bossburg, Krantz became the only scientist to actually follow-up on a Bigfoot report *in situ*. This was indeed a novel experience to any Bigfooter. When Krantz went public with his beliefs that Bossburg was legitimate and Cripple Foot authentic, it was even more novel for the Press. Although an obscure academic with a newly achieved PhD, his endorsement was nevertheless an infusion of respectability (by media standards) into a genre that desperately needed it for its own survival sake. Nineteen-seventy was the threshold of Bigfoot's very marketable decade, a decade brimming with terse news broadcasts and documentaries reaching millions. It needed an expert, and the façade of expertise is always much assisted by brevity.

Popular impression takes on a life of its own after that. Often times the Press can feed it, but just as often the Press can be led by it. It's rather a strange cyclical phenomenon. What it boils down to in the end was best summarized in the irony-climax of the classic western *The Man Who Shot Liberty Valance*, where the newspaper editor, after hearing the buzz-kill confession of a living western legend, threw his notebook into the fire. When the character (played by Jimmy Stewart) expresses his surprise that the editor is foregoing his story, the editor declares the old newspaper maxim: "When truth contradicts the legend, print the legend." How legend gets established is often not important. It has acquired a life of its own and that means it is easy reporting, economic reporting, and, too true, safe reporting.

Exposure of such legends and popular impressions is often then left up to blokes such as me who neither wish to do it but find that in writing a longer investigative thesis than any newspaper could ever sponsor a reporter to undertake we must set the record straight. Years of investigating cold hard facts behind-the-scene of popular impression gives one the weaponry to do so. Facts give one the right. *Right* is, however pure, surrounded by a halo that is gray with fringes that are all too often blurry. Thus *rightfully* so, though courage is not lacking, simple grace often causes one to hesitate.

Grover Krantz's pen may not have written the inexcusable theory that "the most probable" culprit for Bigfoot is *Gigantopithecus*, but he has certainly undersigned it, in doing so giving the image of scientific endorsement to the carnival synthesis of Bigfootery. To look into "The Gentle Giant" *Gigantopithecus* is to look into the heart and soul of Bigfootery. It is to soundly expose its ultimate authors, the unstable and folkloric woodsmen, and to rip the mask off the popular impression that Grover Krantz was a careful and consistent scientific counterbalance to their bizarre investigative methods.

Krantz was, in a very real sense, Bigfoot's greatest victim. He was a latecomer to the pursuit and only knew of the muddled popular image of Bigfoot post 1958 Bluff Creek. There were more than enough of the venal Bigfooters at Bossburg to reinforce the fabulous image of a giant 8 or 9 feet tall. John Green could give him the partial chronicle of this Bigfoot by giving him his recently published *On the Track of the Sasquatch,* in which he makes the equation between the Yeti and Sasquatch as the same thing and then both as the selfsame *Gigantopithecus.* René Dahinden, always eager to find a supporter for the Patterson Film, could quickly agree to let Krantz have a preview. Despite the difficulties between the Patterson camp and his, Dahinden was sure the film was authentic. As a holder of the rights to lecture with it in Canada, Dahinden desperately needed academic endorsement after he showed the film to his club audiences.

With these and a cast of the Cripple Foot in hand, Grover Krantz went back to Pullman, Washington. Once there he could safely ensconce himself in his office at Washington State University and casually examine the data. As the Dickie Davis contract implies, Krantz had deduced rather quickly that Cripple Foot was legitimate. Therefore without *any* historical knowledge on the topic except Green's book showing enlarged human feet at Bluff Creek, Krantz came to his conclusion that Cripple Foot was not incongruous with the old legend of Sasquatch, which Green had declared was one and the same with Bigfoot.

This equation could have been avoided had Krantz not contributed a very unstable factor to the Bigfoot pursuit for which his legacy alone must bear the imputation. He did not believe that anybody was educated enough to fake the kind of evidence that he looked for, at least from what wit he saw in the Bigfooters. Ultimately it was not Green's book that convinced Krantz there was a real phenomenon, it was this blind-sighting presumption. It directed his gaze upon Cripple Foot's "unique bulges." These were tangible evidence he believed beyond a hoaxer's ken, and every other mediating fact was simply shut out. This was evident in 1971, just one year after his Bossburg baptism. He described Cripple Foot for the world on the new documentary *Bigfoot: Man or Beast?* "The foot was twisted and two bulges appear, callused structures on the outside edge of the foot and correspond to gaps in the bone which I've reconstructed here. If this was faked the person doing it had to be an absolute expert in human anatomy."

Paradoxically, however, though Krantz felt his academic position isolated him from the crude woodsmen, he built a creature to conform to the image their gullible brush had painted since 1958. Accordingly, Cripple Foot grew in size until Krantz notably contradicted his own arguments. For *Mysterious Monsters* in 1975 he stated in regard to the same bulges and foot: "If this had been a human foot these bulges would have been farther back. But they are shifted forward making the heel longer and the front of the foot shorter. This is exactly what is required for a foot that's going to carry perhaps an eight hundred pound body. Now, I don't think any faker could have thought of it and figured this out and adjusted his footprint accordingly. A fake footprint is just simply an expanded, enlarged footprint, an enlarged copy of a human foot."

Regrettably, that is exactly what the *uncrippled* left foot of "Cripple Foot" is. Everything about Cripple Foot, in fact, indicated an uninformed hoax. . .except those bulges. But even these underwent a metamorphosis, first supporting a human foot or an expert hoaxer in human anatomy and then denying it in favor of something much larger and inhuman— *Gigantopithecus*. In any case, no matter how the "bulges" could be put in theoretic juxtaposition Krantz believed they indicated too much expertise to indicate anything but a real foot.

The same fluid standard was used in judging the Patterson Film. The contradictory existence of a sagittal crest and breasts on the "Bigfoot" caused zoologists to laugh. But in defending the hermaphrodite qualities of the Patterson Film amalgam, Krantz declared. "The crest, the sagittal crest on the top of the head has been claimed as a male characteristic in the Patterson Film. Well, yes, it is a characteristic of adult male gorillas and orangutans; but it is not a male characteristic, it is a size characteristic. Beyond a certain size the jaw muscles must find attachment on a special crest. And it happens that only male gorillas and orangutans get that big. If there were a female primate of the 500 pound body size it would have to be so big that it would have a crest as well. So it's not a male characteristic."

Zoologists did not support such morphological claims.[40] A combination of both characteristics required disputing one or the other if Patterson's clumsy amalgam was to survive. The crest was undeniable, but Krantz challenged the breasts. "In Patterson's film the creature has a bulge here which seems to be a two part bulge, but it is not really very clear. This might not actually be breasts. These could be out-pouchings of the trachea which occurs on many large apes called laryngeal air sacs." Perhaps this argument is forgivable in the 1970s as he crystallized his theory, but it is also repeated in his 1992 book *BIG Foot-Prints.* By this time it was hopelessly antiquated, as even most Patterson protagonists accepted those were obviously breasts. Krantz, too, felt that breasts were the more probable. "Actually I do not find this to be as likely an explanation as that they are breasts. I just think it is good practice to look at every reasonable possibility. . ." thereby leaving him with nothing else but his unconvincing arguments about the evolution of a sagittal crest as a size characteristic.

After 22 years if still questioning the sex of the subject in the Patterson Film is possible, then surely the evidence presented in it is not very clear or not conducive to the investigator's desired theory. By 1975 Krantz was able to admit that "I've checked just about every possible measurement of that film, having looked at it, oh, at least 50 times now." The reason for accepting the film, however, seems more an echo of his profoundly disturbing standard. "I don't think there is

---

[40] Besides, Roe was explicit: the female he saw had no pointed head.

any way that Patterson could have figured this out to have faked it."
Academic, however, is not intelligence. I prefer the words of Dr.
Frank McLynn who, holding the chair of history at Strathclyde University, Glasgow, was partial to remind his students that "academic and intelligent are very far from matching compliments." The same can be said for ingenuity. Hoaxers have shown they have great ingenuity, and equating this characteristic with requiring a scholarly background is absurd.

Shortly after the publication of Krantz's book *BIG Foot-Prints* this was sadly revealed. The sacrosance of his maxim "I cannot be fooled by fakers" was shattered in a dynamic way. *Skeptical Enquirer* contributor Michael R. Dennett was actually able to track down the faker of the Indiana Print that Krantz had touted in his book as the best proof for a Sasquatch existing east of the Rockies. The faker, J.W. Parker, admitted how he used his hand skin to impress dermal ridges. He also used a walnut shell to make the big toe look more interesting. He then made the cast and mailed it to Krantz.

To add injury to insult, Krantz's "secret" criterion to determine authenticity became publicized. He had always said that he kept two vital pieces of evidence from the public because these two helped him determine fakes from real prints. He had lauded the Indiana Print as containing these two foolproof pieces of evidence. Dennett asked Parker what he did. Parker thought a moment and then recalled he put in scars and toenails.

Parker had made the imprint in 20 minutes, even depressing the ball of the "foot" to simulate the push-off of a foot in walking stride.

The exposure of the Indiana Print capped-off a decade in which determining "real" Bigfoot evidence depended on Krantz's pronouncement of finding "dermal ridges"— skin detail— in plaster foot casts. This, too, was more evidence he thought beyond hoaxers' skill. "It would take an expert at dermotoglyphics to reproduce absolutely perfect skin detail." Sometimes these dermal ridges amounted to no more than a small patch on an otherwise muddled plaster cast.

Thanks to this standard Paul Freeman had dominated the 1980s as the man finding the most Bigfoot evidence, sometimes just while casually going about his routine patrol in the Blue Mountains of Washington. By 1992 he had cast more than 30 footprints, more than any original Bigfooter. Given how supposedly rare Bigfoot was, this seemed suspicious to many Bigfooters. Equally suspicious was it

Indiana Print

that Freeman should be the first to find them with so many dermal ridges. Freeman even preempted Krantz's study of the prints, naming one of the Bigfeet "Dermals" in honor of all the skin ridges his casts showed (before handing it to Krantz). There seems to be no doubt that Freeman knew that Krantz looked for those as a point of authenticity. When Freeman produced hand casts, Krantz admitted that he had loaned Freeman a hand cast that none other than Ivan Marx had given him long before— not a pedigree of great caliber.

Freeman even cast knuckle prints, which, of course, accentuates ape knuckle walking. Whoever made these wasn't aware Sasquatch never goes down on all fours. Although *loysi* is said to, this could not be from a *loysi* as there is a thumb. It also can't be from a Skoocoom since the cast shows an opposable thumb.[41]

Naturally, Krantz's high media profile added weight to his public pronouncements of a cast as authentic. This led to a lot of enmity between him and his prime Bigfooter detractor, René Dahinden, who remained adamant about these Freeman Blue Mountain/Mill Creek prints. "Look, any village idiot can see that the tracks are fakes!" At one point, in fact, the footprints went in echelon: right, left, right, left, left. . .leading any observant investigator to question them. Krantz could also accept that hair left at some of these track sites was synthetic and yet accepted the tracks based on his belief nobody could have faked areas of skin ridges on them. "Every time you open your mouth to the Press," Dahinden told Krantz, "you make a bunch of downright stupid statements."

Dahinden's outburst only minutely reflects the rumblings that existed in some quarters of Bigfootery against Krantz's high profile and yet apparent lack of any real coherent historical or field investigation. Dennett quotes one detractor of Krantz, Cliff Crook, who scathingly said: "Science is about discovering the truth. It is evident that Grover Krantz has consistently abused his scientific credentials by his constant failure to acknowledge plain facts."

More than any critic could, *BIG Foot-Prints* perhaps revealed a very uncharitable image of Dr. Krantz's actual level of valuable input in the topic. By his own admittance Parker's plaster cast had merely been mailed to him. The others in his armory had been handed to him. By his own choice then he had placed actual active on-the-spot investigation second or, more often, nonexistent. On top

---

[41] Krantz denies this. He says of the picture in his books that it is a nonopposed thumb. A nonopposed thumb would flex and bend with the fingers, not stick out along the side. The thumb is clearly opposable.

of this his own method was highly questionable. As early as 1974 he had told Robert Guenette (for the making of *Mysterious Monsters*) that he "has no special criteria in determining if a person is telling him the truth. He trusts rather to his instinct in sizing a person up . . . in comparison to other known facts."

Hindsight shows, however, that by 1974 Krantz had never seen any legitimate Sasquatch evidence. His thesis was only based upon examination of the Cripple Foot cast and accepting the Patterson Film as genuine. Thus the "known facts" which anchored his theory and influenced his interpretation of incoming data were completely false and anti-theoretical. Consequently, his "instinctive" criteria led to him dismissing all historical data, witness depositions and circumstantial evidence as hearsay that disagreed with these "known facts." In the process he dangerously isolated his two pillars of evidence from the body of work that openly contradicted their legitimacy.

In *BIG Foot-Prints* Krantz admitted that he was as naïve as anybody could be when he entered the fray. This is completely forgivable. But after 22 years he had had more than enough time to expose Bigfooter concoctions and set the record straight. His book's publication was anticipated as the scientific expert speaking out in detail after so long in the pursuit. However, *BIG Foot-Prints* surprised and disappointed many by how it revealed the paucity of Krantz's actual progression on the subject. Cripple Foot and the Patterson Film had remained the cornerstones of his evidence, and his arguments remained the same as those in his spots in old 1970s' documentaries. Upon an enlarged human foot and the Patterson amalgam Krantz continued to endorse the Bigfooter theory.

"As far as I can determine," wrote Krantz, "it was Bernard Heuvelmans (1952) who first suggested the connection between *Gigantopithecus* and the Himalayan yeti. Ivan Sanderson made the same proposal in 1961, he was soon joined by anthropologist Carleton Coon (1962), and many others have since agreed. In 1968, John Green made the specific equation with our North American sasquatch, and this has become the near consensus of opinion." Furthermore, "I made the contention in a publication in 1986 that we in fact have footprints from *Gigantopithecus blacki* here in North America." Later, "My argument, much like John Green's, is based on the most probable interpretation of *Gigantopithecus* as a bipedal hominid that has no other human traits, and the same interpretation can be made for the sasquatch from footprint evidence alone. To propose that two different kinds of animals are involved, with essentially the same description, would be unlikely in the extreme. Since there is only a small amount of tangible evidence from either source, and no direct overlap between them in what material there is, it must remain a possibility that there is some degree of difference between these Asian and American giant primates. . .It is possible that there might come to be some degree of acceptance for the name *Gigantopithecus blacki* for the sasquatch. . ." Lastly, "I would give this species a distribution that probably encompasses much of North America, and most likely includes some of northern Asia."

Assertions such as these could not come from anyone with even an iota of knowledge on the Yeti. Because of this, John Green's *Gigantopithecus* was not simply the misplacing of a fossilized prehistoric Eurasian species in time and in America. It was a bastardization of the real thing. Yeti, for all of its enigma, did (does) have a peculiar "apeman" type of foot and a general and consistent description that fits that of an unknown anthropoid. Those who speculated it could be *Gigantopithecus* were on somewhat surer ground and at the juncture of forgivable error if proven wrong. But if *Gigantopithecus* is also the owner of the Cripple Foot, a *human* foot, then what is it? Then what is Yeti? For those who had followed Bigfoot and Yeti throughout these decades, Grover Krantz's road to *Gigantopithecus* proved he had not even a passing acquaintance with the basics of international crypto-zoology.

As a physical anthropologist Krantz was completely unqualified by academic training to tackle the questions of actual fieldwork in zoology; and despite his book's promo that he "is a world authority on crypto-zoology" it was clear he didn't have *any* working knowledge of either Yeti, Almas, or actual Sasquatch data.

A comment in *BIG Foot-Prints* underscores Krantz's limited knowledge. Of the most studied foot of the 20th century, Krantz wrote of the Shipton Photo and footprint: "Ivan Sanderson made much of this photo, along with native reports, to build a picture of some kind of ape that was only partially bipedal. Until another photograph or cast of that particular foot design turns up, I see no point in taking it seriously."

Upon that foot Heuvelmans, Tschernesky,[42] Coon, and so many others *had built* Yeti and speculated on *Gigantopithecus*. None broached that Yeti and Sasquatch were the same thing. Only John Green made the connection. What impelled Krantz to ignore so much data in Izzard's, Stonor's, Heuvelmans' and Sanderson's books and follow him?

[42] Tschernesky was actually first, not Heuvelmans as Krantz erroneously thought.

There is probably more than one reason. But in *BIG Foot-prints* Krantz perhaps unwittingly revealed a terrible conflict of interest that undermined his ability to objectively examine and compare data. "It is difficult enough to get most of my skeptical colleagues to listen to anything about the sasquatch evidence, and almost impossible to make them take more than a cursory glance at a footprint cast. Still, some of them are curious enough to want to know a little more about the subject, and I am the person they often come to. If I were now to assert that there are multiple species of hairy bipeds out there that have not been demonstrated, I would lose what little credibility I have now. . .Fortunately for my own credibility, I do not see any compelling evidence for more than one type of hairy biped."

Yet the footprints from Krantz's 2 pillars— the Patterson Film subject and Cripple Foot— were so radically different that even John Napier, himself an anthropologist, had noted in *Bigfoot* (1972) that if both were real, *always if,* then it indicated more than one species was behind Bigfoot itself.

Furthermore, several tracks of Yeti footprints had been uncovered by the *Daily Mail* expedition showing the Shipton Print was not a one-time encounter. Then in 1972 the McNeely Print was found and this should suggest to the informed that it is the Almas foot. Krantz was either unaware of these or dismissed them because he did not personally examine any casts of them. By 1992 he was still completely unaware of 4-toed "Skoocoom" prints and even offers his surprise when he saw the Manitoba 1988 prints, the only real prints he ever saw. He had no knowledge that the Skoocooms had 4 toes or of the image of *Ameranthropoides loysi* reflected in so much Indian artwork.

To the uninitiated in the quirks of anthropology it may appear as if Krantz's unique investigative criteria led to his belief that human foot tracks in the wild could indicate an ape. In this area, however,

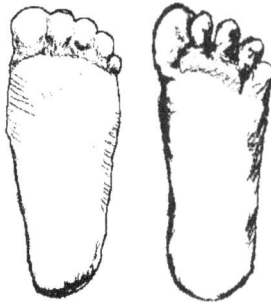

*A fatal comparison: Patterson's Bigfoot and Cripple Foot's uncrippled left feet.*

he was following the lead of a small group of anthropologists. At one point in time elements of uniformitarian paleoanthropology asserted that *Australopithecus* essentially had human feet and walked upright just like a man. One of the most humorous pieces of evidence used to back this notion were the human footprints found in volcanic ash at Laetoli, Tanzania, by Donald Johanson, an anthropologist given to sudden and sweeping inductive comments. During the 1970s in an era in which Uniformitarianism was just cracking wide open, Johanson popularized the idea that these footprints were 3.6 million years old (using uniformitarian dating methods of Day-Age). By the same mentality of sedimentation, it was popular to believe that Australopitheci were as old as this as well. Thus, in examining them, he declared: "The footprints would have to be from A. afarensis [species of Australopithecus]. They substantiate our idea that bipedalism occurred very early, and our contention that the brain was too small to master tools."

The Laetoli track of prints shows that a purely modern human foot is walking over the ash. Mary Leakey, Richard Leakey's mother and wife of Louis Leakey who popularized Australopitheci, even later noted that a small child seems to be intentionally lengthening its stride to step inside the print of the adult. From this evidence the prints sound suspiciously modern. African Bushmen are known to start training their children at 5 years of age to hunt. In all probability, the Laetoli footprints reflect a track left in the last 5 thousand years by some African Bushman and his son. Many other prints of modern animals were preserved in the ash amidst these prints as well. There was no evidence that indicated these prints were prehistoric.

The only way such a ridiculously long age was given the volcanic ash was from the now-defunct Geologic Column concept. Potassium-Argon and Argon-Argon ($Ar^{40}/Ar^{39}$) gave such widely disparate dates that per usual during Uniformitarianism the standard practice was to defer to the Geologic Column concept to settle the issue and calibrate the apparent date by assuming a very slow process rate of the rock.

Radiometric dating, such as the aforementioned Potassium-Argon, has fallen into deserved disrepute. Examples of erroneous dates include the famous 1801 Hualalei, Hawaii, flow which yielded dates of anywhere from 160 million to 3 billion years. Kilauea basalts, also about 200 years old, dated to 22 million years. Potassium has proven too "mobile" to ever contribute to accurate dating. Using distilled water, as much as 80 percent of potassium can (and has been) removed from small meteorite samples in less than 4.5 hours.

Paleoanthropology, the study of hominid bones, is itself sometimes a willy discipline that is questioned by its chief practitioners. David Pilbeam, Harvard's paleoanthropologist, wrote in the forward of Richard Leakey's 1978 *Origins*: "My reservations concern not so much this book but the whole subject and methodology of paleoanthropology. But introductory books— or book reviews— are hardly the place to argue that perhaps generations of students of human evolution, including myself, have been flailing about in the dark: that our data base is too sparse, too slippery, for it to be able to mold our theories. Rather the theories are more statements about us and ideology than about the past. Paleoanthropology reveals more about how humans view themselves than it does about how humans came about. But that is heresy."

Few would endorse these Laetoli footprints as proof that they belong to *Australopithecus* and therewith that *Australopithecus* had perfectly modern human feet and walked just like a man. Most may not agree with the great anatomist Solly Lord Zuckerman's emphatic "They're just bloody apes!", for we have already seen how detailed study indicates a very different locomotion for *Australopithecus* than humans and apes— something that, given their size at around 6 foot at best, is a little more interesting to the topic of the real Sasquatch and more so to the Almas.

By 1992, however, Krantz must still have believed that *Australopithecus* had modern human feet. This more than anything must also have underpinned his belief that an "ape" walking just like a man in the Patterson Film could be real. He so much as said his *Gigantopithecus* could be a giant *Australopithecus*. His other methods for determining that *Gigantopithecus* walked upright were more original to himself. Because the last specimen of *Gigantopithecus* was dated to 300,000 years ago (by Uniformitarian methods), Krantz essentially believed that Sasquatch in turn substantiated that *Gigantopithecus* was bipedal because 300,000 years is not enough time to affect great evolution (by Gradualistic Theory).

Perhaps because *Gigantopithecus* is a scientific name it has lent, however ignobly to the informed, the image of being a relevant theory to the rest of the hearsay population. In actuality, John Green arrived at it purely by folklore and hoaxes, both of which stretched Sasquatch into a giant ape. Green's equation, however, was merely based on the existence of a large primate jaw found in India and on Tschernesky and Heuvelmans' belief that Yeti walked upright and was a giant. None of it was based on any description of the actual

[ 271 ]

6.6 foot tall, small-headed Canadian Sasquatch.

Indeed, it seems that Green had absolutely no knowledge of Sasquatch. Despite J.W. Burns' stories, which made it clear the Sasquatch was hairy all over like an apeman or animal, Green went to press as late as 1968 with his first book, writing in *On the Track of the Sasquatch* that "They were called hairy giants it is true. But this was taken to mean that they had long hair on their heads, something along the lines of today's hippies." This really is John Green's individual assumption and not the collective assumption of White Man. Nor is it anything that could have been gleaned from even a cursory reading of Burns.

Green was not a native of the Saskahaua, this is true; but after living there so long it is ironic that he remained that uninformed so that this comment could be placed in his first chapter "Meet the Sasquatch." Along with this there is a worse misrepresentation of J.W. Burns. "Burns got his stories from his Indian friends, and some of them smacked heavily of the supernatural. Thus Burns' deserves not only the credit for introducing the Sasquatch to the general public, but also the blame for affixing firmly to it the label of Indian Legend."

Without such an unjust divorce from the original image Burns compiled of the 6 to 7 foot tall Sasquatch man, *Gigantopithecus* could not have been introduced and substituted into its place. Popular press, not Burns, had changed Sasquatch into an 8 foot tall giant during the 1930s (even in some of the same vignettes Burns had used in his articles where they were 6.6 feet tall!).

Throughout the clumsy life of Tom Slick's venerable PNE (in concept), it was clear its failure was predestined because Slick concentrated in the land of Bluff Creek's bogus Bigfoot. Had Green understood he himself lived at the fringe of the Saskahaua, the very heart and soul of it all, he could have pled a very convincing and passionate argument to send an expedition to Morris Mountain. Yet 4 years after PNE's birth, Slick was still without direction. If S. Kirk Johnson was becoming fed-up with some of the Bigfooters' input he had good cause.

From René Dahinden's account of PNE's activities (in *Sasquatch*, 1973), it seems clear there was much wanting in the original Bigfooters, especially Bob Titmus. Yet unwittingly Dahinden revealed more negative about himself. Much of the tenor of *Sasquatch* seems to be Dahinden settling an old score and shouting his superior dedication. Yet it was hypocritical of Dahinden to constantly denigrate and insult the other Bigfooters. Their "antics" which he "judged with his forthright criticism" did not cause PNE to fail any more than his incompetent input and poor analysis, both of which led him to believe the Bluff Creek hoax was legitimate. With his crowing of being the first to uncover that Sasquatch was really a Yeti— thereby trumping them all— he becomes the godfather to the greatest error that led to *all* other errors. And this began with nothing more analytical than a talk with his milk ranch boss.

As a new immigrant to Canada (in Calgary) he heard the announcement of the *Daily Mail* Expedition on December 3, 1953. His boss, Wilbur Willick, was with him. "Now, wouldn't that be something," declared Dahinden to Willick, "to be on the hunt for that thing?" Willick, who had worked on the West Coast, responded: "Hell, you don't have to go that far: they got them things in British Columbia."

Dahinden admitted that something had "clicked" in him. He got Willick to elaborate and soon discovered that the Indians had stories of hairy giants. After Dahinden moved to British Columbia in 1954, he couldn't shake the thought. He started asking around, receiving in reply that it was all an Indian myth. However, he did extensive research in newspaper archives and the Vancouver library. After uncovering old stories, he came to the conclusion there was indeed something to find.

From day one Dahinden's glasses were Yeti-tinted. This did not change even after he met and eventually became friends and working partners with John Green, who would secure a tracing of the Ruby Creek Print which proves Yeti and Sasquatch are radically different. (Dahinden even worked for Green at his newspaper for a short while.) Ironically, one of his book's purposes was to show how he antedated Green in the pursuit of Sasquatch. He even takes care to supplant John Green's reputation for this honor by mentioning his first meeting with Green in 1956, a meeting in which Green told Dahinden to forget about the Sasquatch myth. With his arsenal of research it seems that Dahinden was indeed quite an influence on Green after Green started to believe. Indeed, Dahinden might have proved a greater influence on swaying Green to believe than did Jeannie Chapman's story and that of William Roe.

And John Green would prove to be quite an influence on Tom Slick. Dahinden, through Green's new "complete enthusiasm," might even be the inspiration for PNE. On October 15, 1959, the *Humboldt Times* announced its founding. Heading: SLICK THINKS BIGFOOT KIN OF ABOMINABLE SNOWMAN. When Dahinden finally joined his buddy Green in November as a member of PNE he doesn't seem to have contributed anything to it but its failure. His chortling criticisms of the others are contemptible when he musingly recounts how he found no use for the brownie box cameras watching baited areas so he arbitrarily removed all the film and loaded it in his own camera. He didn't tell the others. He just sniggered because Green and Titmus hadn't noticed. They continued to lay bait, not knowing there was really nothing on the other end that could take a picture when the string was pulled. Dahinden admits he could only stand a couple months of PNE (November-December 1959) and then left. He returned in January 1960 and staid only a short time, seeing it as nothing but a comedy of errors. His biggest assignment was to actually go with a "sheep herder" who was also some "half-ass poet" and recover a couple of donkeys that S. Kirk Johnson Jr. abandoned in the wilds after he felt the need for high life in Orleans, California. Dahinden and this sheep herding half-ass poet whose name he never bothered to learn were dropped in by helicopter and a couple of days later, boozed to the gills, found their way out by following the sickest looking donkey.

Dahinden's analytical inabilities truly flowered after a trip to Russia in 1972. There he interviewed Porshnev. Dahinden was sure that the Almas and Sasquatch were identical. Porshnev also regaled him with stories of Neanderthals. Remember Bigfoot had human feet! After he returned from Russia he remained on a high for years. In 1974 he adamantly declared of the Bigfoot pursuit: "You have to be cynical. You have to be rutless— It's no game! It's no game! It's one of the biggest scientific discoveries the world has ever seen. . .in an anthropological and probably zoological and biological sense. It might rewamp or turn over the whole theory of evolution and everything else. . .it will flip our lid, to put it bluntly. So therefore never mind the philosophical discussions. I say go krab it any vhich vay. I'm in dis to catch the Sasqvatch dead or alive to prove its existence and that's all there is to it!"

Most, if not all, of Dahinden's life was spent working at odd jobs (like his short tenure with PNE) or starting minor businesses (like a boat rental) that later went fluey. Frequently this was because of his interest in Sasquatch. He admitted it became an obsession and that it would control his life. He took a menial job at the Vancouver Gun Club in 1963 (picking up the spent led) just so he could devote all his time and energy to "tinking" about Sasqvatch. "I didn't get sidetracked by a vife, kids, and cats and dogs and mortgages and all this *crap*." He sacrificed his family, any possible career he could have had, and his friends.

[ 274 ]

In Dahinden's driven agenda to preserve in print his actual foundational role, history now reveals a Catch 22. His book has for posterity's sake unintentionally become his confession as the author of the phantom that beguiled the Bigfooters. His own creation— the cone-headed giant, part Neanderthal, part Patterson Film amalgam— would elude him all his life. In his interview with Greg Long he even discouraged searching for Bigfoot. "Anybody who goes looking for Sasqvatch is stupid because the chances of finding something are so remote that you just shake your head."

This was a complete about face to his own statement after Bossburg in which he said he had examined 3,000 prints in the brush and he was convinced all were genuine— a number that invalidates his above statement on the rarity of Sasquatch prints to Long some 30 years later. With Bigfoot's popularity waning over these decades, naturally enough so did the number of tracks. Dahinden's remark reveals he must have finally suspected hoaxing had underpinned much of Old Bigfootery. In fact, in later life Dahinden was even referred to as a "man who investigated men who investigated Bigfoot rather than an actual Bigfoot investigator."

Despite Dahinden's condemnation of Green as "naïve," "selfish," and "shallow," there was little difference in how the two mixed and thumbed the data. Green's prolific output of Bigfoot stories actually undermined his own theory and showed, more on the negative side, how both he and Dahinden (who believed the same stories) made little analysis of their contents. He had catalogued so many diverse stories of the "animal humans" that his equation of *Gigantopithecus* at the end of his book for Sasquatch was incongruous, to say the least. In spite of being opposite poles, they were identical in their analytical abilities. Both mixed and matched and seemed enthralled at their own Frankenstein needlework. Together they became the creators of the Gentle Giant. Their Frankenstein differed from the mad Baron only in that neither could get all the pieces to fit.

This was evident during the Bossburg fiasco, from which Green was still feeling the public scorn. In between his attempt at an ambidextrous juxtaposition with both camps in the feud, Green had committed the public bonar of declaring that any Sasquatch that crosses his rifle scope "would be a dead Sasquatch." There was a public outcry that seemed justified, considering Green had been the conduit for such stories as Albert Ostman's that spoke of Sasquatch being an early caveman grunting "Soka, soka" and "ook."

More than at any time now Green needed endorsement for his

*Gigantopithecus*, purely animal and no more. The Patterson Film, which Krantz was now studying at his office at Washington State University, might clinch the association. When Krantz agreed, he became an immediate object of referral by Green to the Press, and this no doubt secured Krantz's crucial and influential place in the early and pivotal documentaries, *Bigfoot: Man or Beast?*, 1971, and the granddaddy *Mysterious Monsters*, 1975, and from there to many magazine articles and more documentaries. Green had made possible the theory of *Gigantopithecus* and secured Grover Krantz's placement as the one the media would go to find scientific support for it. The irony is obvious— Green had no real modern data; Krantz only had Green's data, but by virtue of his academic position he became the beneficiary of the Press image of "expert" in the new era of Bigfoot popularity.

The significance of any further criticism must therefore bear in mind it is attacking a public media image as a "Sasquatch expert," one which Krantz really was not responsible for and even often lamented. Media forums elevated to a more concrete level what were really often fluid theories that emerged in his very first year of Sasquatch study (1970) and condensed, often with great convolution, only when attacked. If he didn't advance much beyond this seminal year it was really due to the truth of his often underrated statement that he merely "did this on the side."

It is in the light of haphazard and sporadic interest that one can best understand and explain why he, though an honest gentleman, was nevertheless the source of remarkable error when it came to the details of almost any Sasquatch report. This too was sadly revealed in *BIG Foot-Prints*. For Albert Ostman's encounter with 4 "sasquatches" in 1924 Krantz wrote that "his description of them agrees with that of other observers. . ." when in fact Ostman's was notable for being enormously at odds with others, especially in his description of the "mother" having wide hips and all of them having feet with huge big toes and being capable of speaking "soka, soka" and "ook." Krantz's next faux pas is to state categorically that "Sasquatch footprints are roughly human in design. . .Initially it was their great size and often the depth of the impression that set them apart from human footprints. . ." And then, amazingly, he adds: "Further observations include near equal-sized toes that lined up almost straight across. . ." which is exactly the opposite, chronologically speaking.

Even Krantz's recital of contemporary evidence was in error. For Cripple Foot's circumstances he declares very inaccurately in his

*BIG Foot-prints*: "When Cripple Foot left his longest trail back in early 1970, the tracks came out of a large lake and eventually went back into it." Actually, it was late 1969, and the tracks ended at a river. Perhaps Krantz could be referring to Marx's 1970 claim that he lost his wounded bow-legged Bigfoot at Beaver Lake after filming it— if so it conjures up a frightening picture of Ivan Marx's private influence on him.

In fact, Krantz's dependence on Marx underscores the instability of his "no criteria approach" to the whole topic. Blind reliance on his ability to see through any faked evidence perhaps made him think the quality of the witnesses and the circumstances were of less importance, in Marx's case obviously of no importance.

Disregard for the importance of actual on-the-spot evidence is all the more remarkable in light of Dr. Krantz's sensational views. There was an unusual vignette on page 6 of *BIG Foot-Prints* that is a good example of this. He relates the circumstances of his wife possibly seeing a Bigfoot while daytime driving in Washington State. "Only after several miles of travel, while carefully eliminating from her mind all the possible explanations, did she tell me what happened." The circumstances would still seem sufficient for him to deem it necessary that they turn around and check for evidence. Tracks, if found, would certainly be hot and traceable. Yet they continue to drive on, a rather interesting example of investigative ennui.

The image of a dedicated Bigfoot hunter was also smashed with the release of *BIG Foot-Prints*. "People often ask if they can go on my next 'expedition' to hunt for Sasquatch. I don't go on expeditions, nor do most of the hunters, at least not in the usual sense of the word. All of us make frequent trips, either to a specific goal or just to scout some territory. Rarely do such trips involve an overnight in the bush. Given the presence of Sasquatch right on the edge of civilization, and the fair network of roads in much of their territory, there is little reason to go to a lot of trouble to search for them somewhere else."

Only the garish world of Bigfoot could have caused one to believe Sasquatch frequently walks across our highways today. This same world seduced Krantz to outlandish comments. From all the reports, Krantz deduced that as many as 10,000 sasquatches exist, about twice that of orangutans in the wilds. Though a very endangered species, they are studied and followed consistently. No discovery has ever come of Krantz's giant Sasquatch.

Over his life Krantz had come to feel his Sasquatch research had damaged his career. "My first shock was to be deferred on tenure for one year, in spite of the fact that I was already more frequently cited for my normal work than most of the other members of the department. Some lame excuses for this deferral filtered back to me, but the media coverage of my Sasquatch work was a more likely reason. Much later, when my promotion to full professor was denied, the only reason the dean could come up with was that my work was not favorably received in all quarters. . .Though it was quite specifically denied to my face, I have no doubt that my sasquatch work was the major deciding factor against promotion."

This was unnecessarily unjust on Krantz's part. From how he incongruously mixed his Sasquatch theorizing I can imagine peer reviewers of his normal work were sometimes left disconcerted.

Superiors and colleagues at Washington State University had actually accused Krantz of "fringe science" in his approach to the Sasquatch question. After his death in 2002, his superior Bill Lipe, professor emeritus at WSU, said: "Most of us didn't agree with him that his evidence was very good." Lipe insisted, however, that this was not what kept Krantz from being promoted to full professor. "He couldn't publish his articles on Bigfoot in peer-reviewed journals, and he didn't seek the research grants. . .The evidence never got any better." Donald Tyler, Professor of Anthropology at the University of Idaho in Moscow, said in regards to the cornerstones of Krantz's evidence: "He thought the evidence couldn't have been faked. I sometimes think he was too smart for his own good."

Didn't seek research grants? That is indeed hard to understand in view of Krantz's own statements in *BIG Foot-prints*. "There is an arrangement with the Washington State University Foundation whereby research contributions to my university can be earmarked especially for sasquatch investigation. This was set up back in 1986, and no donations of any kind have been received for this work." He also received no donations from a display of Cripple Foot in the San Francisco International Airport art exhibit— "The Right Foot"— soliciting investment. He also achieved nothing from being the nominal president of a tax free organization, the Northwest Hominid Research in Pullman, Washington.

There is no denying that Krantz's claims of his own involvement on the topic varied and were sometimes also inconsistent, but, with his own impoverished evidence at hand to bear witness, it seems safe to tip the scales to the side of lackadaisical interest.

Some may view this as avoiding the issue or talking around the real underlying problems and very disturbing elements that led Krantz to his conclusions and to refuse to relinquish them. Be that as it may, it is a better row to hoe than acrimony. It is true that Krantz's peculiar method of investigation made him prone to the carnival of Bigfootery. It is also true that his encounters with Bigfooters also gave him a false sense of security against hoaxers, for he certainly declared none could fool him; but it should be with sadness and irony that one looks back on the lifetime of paltry evidence he accumulated rather than acrimony. He became a Falstaff to a carnival king, and though this is not a laudable position it is also not one that should be the butt of mockery. It is more important to heap earth upon *Gigantopithecus* rather than to heap stones upon Grover Krantz. Many a kid from the 1970s, this author included, took comfort in Krantz's arguments in favor of Bigfoot.

Institutionally, Science has had to confront individual members who became dangerously enamored of some piece of evidence or theory of which they would never let go. Anthropology is particularly prone to this. Stephen Jay Gould would remind his readers that anthropology is such that "experts" find what they are looking for regardless if it is present. "Information always reaches us through the strong filters of culture, hope, and expectation." In using the notorious hoax of Piltdown Man for an example, he says of Sir Arthur Keith's examination of it: " 'His forehead was like that of an orang, devoid of a supraorbital torus; in its modeling his frontal bone presented many points of resemblance to that of the orang of Borneo and Sumatra'. . .Careful examination of the jaw also revealed a set of remarkably human features for such an apish jaw (beyond the forged wear of the teeth). Sir Arthur Keith repeatedly emphasized, for example, that the teeth were inserted into the jaw in a human, rather than a simian, fashion."

Sir Arthur Keith spent much of his life studying the puny remains of Piltdown Man. Before he died he was able to see the entire hoax exploded by chemical analysis in 1953. Nothing more than a modern human skull cap and an orang's jaw with a patina to make it look ancient was this the great missing link. Keith was the victim of his own limited standard. He declared: "The only kind of being whose existence is testified to by scores of witnesses, and which never reaches the dissecting table, belongs to the world of spirits." In similar attitude, Dr. Grover Krantz continued over half his life creating Sasquatch from a few plaster casts of enlarged human feet and a 38 second strip of film while ignoring all witness testimony, contemporary and historical data.

Of Krantz's evidence we can say the same thing that Dr. Heuvelmans said of Keith's: "Before he died in 1955 Sir Arthur Keith had an opportunity to see the folly of trusting too much to 'concrete and unmistakable evidence.' Had he taken more notice of the circumstantial evidence he would have realized sooner, as many other anthropologists had done, that the Piltdown Man was morphologically and chronologically inconsistent." Time and chronology are equally devastating to Krantz's *Gigantopithecus.*

In conclusion we should defer to Solly Lord Zuckerman when he said that there really is no science in the search for man's fossil ancestors. It starts at physics and ends in the field of "presumed biological science" where, declares the noble lord, "the ardent believer is sometimes able to believe several contradictory things at the same time." Fulfilling Lord Zuckerman's comment, the old Bigfooters could proffer the attitude of big game hunters out to shoot a specimen while at the same time fondling (and promoting) the idea they might have a missing link as their quarry. Such frightful contradictions are hardly born in the thinking mind. But perhaps the thought that a missing link was involved added to the glory of obtaining the prize. It certainly added to the legend of Bigfoot we grew up with.

When all is said and done should we believe in the Gentle Giant? None of its chief protagonists could even distinguish between the Ruby Creek Print and the Shipton Yeti footprint. In between these two radically different footprints they built a striding giant that landed in America with enlarged human feet.

CHAPTER 15

## Saskahaua Enigma

An enigma is born from facts that do not tally up into one under-standable whole. For years the Saskahaua provided just such an enigma— wild men, "animal humans," beast men. Saskahaua Georges were something man and they were something animal. They were two tribes, obscure, vague, seemingly mythical.

In the hubbub of Bigfoot popularity, theorizing wiped this enigma out and gave us *Gigantopithecus*, solely indivisible and individual, a prehistoric Eurasian giant ape now become an American bipedal missing link. We replaced an enigma with an improbable mystery. I will make no small fuss out of the difference. A mystery is something unknown or unproven. An enigma is something not understandable.

The enigma of the Saskahaua is very legitimate. The maker of the strange Ruby Creek Print is real, sometimes even attaining a height that reflects the image of the legend. What is not real is the object of the legend. But the quest is legitimate. The early image of Bigfootery is an image to aim for. Today it must replace the cone-headed giant with the enigma. A new trail is waiting to be blazed. Its object must come from the world of the monkeys. We stand on an exciting threshold again, just where the Bigfooters stood 52 years ago. Just as they had pondered whether humans were involved, so must we dis-cover if people are truly involved. The words of Chief Flying Eagle still apply: "The white speaker is wrong! To all who now hear I say: Some white men have seen Sasquatch. Many Indians have seen them and spoken to them. 'Sasquatch' still live around here. . ."

The San Francisco *Chronicle* gave us an interesting comparison 46 years ago in 1965. Except for the height of the "Man Animal of the Northwest" could it reflect a true image of one of the entities we still seek?

Dahinden's observation of the Bossburg Incident has sadly proven prophetic for the entire search for Bigfoot. "The longer they went without finding a trace, the worse the suspicions grew" that it was all phony. It's 52 years now since Bigfoot reality in 1958, and need I say that nothing has been concretely found. Quite the opposite has happened. Bigfootery erased the original Sasquatch footprint; obliterated the Indian testimony of something 6.6 feet high and it also glossed over and forgot that more than "one tribe" was involved. Bigfootery has revealed nothing to us but the reality of pure California folklore, not the track of the true Sasquatch, the "animal human," but the track of the Bigfoot, the publicity hunter. Real Sasquatch reality has vied only weakly for recognition amidst all the hype.

And now that many no longer believe in the mythological Bigfoot, it becomes easier and vital to discard it and establish the reality of the Sasquatch and the Sasquatch Man beyond all doubt.

It is of the utmost importance that a specimen be captured or, at the very least, filmed for a prolonged period of time. Nothing else will do. The idea that killing one will prove all we need to know is lame. It will not prove to the majority of scientists that 'it' or 'they' habitually walked upright. This would require the study of a living specimen. There is too much that needs to be studied concerning the Sasquatch, the Almas, Skoocoom or the *loysi*, and a corpse cannot answer all the questions.

To this end an expedition, a real scientific expedition, is essential. Its objective must be one place. It's time we started regarding the facts we have and go from there, go to real Sasquatch country, to a land that is still unknown.

"Peter remarked," wrote J.W. Burns in 1929, "that his father and numbers of old Indians knew that wild men lived in caves in the mountains— he had often seen them. He wished to make it clear that these creatures were in no wise related to the Indians. He believes there are a few of them living at present in the mountains near Agassiz. . . They may live in other places throughout the province," said Charley Victor, "but I have never heard of it."

# The Mountain Giants

*A more accurate comparison?*

Other sightings have proven Brave Charley Victor wrong. The mountainous Saskahaua is only the center, the navel of the vast *terra incognita*, the great unknown and hidden land of the Pacific Northwest. The crown is the great and mysterious Mackenzies and Selwyns and the Nahanni Valley nestled in their bosom. They are so remote they still hold a fabulous but true legend of a lost gold mine (the McLeod Lode), a semi-tropical valley, and the truth of gruesome discoveries that may help direct us.

The publication of the Fortean Society of New York, begun by famed writers Ben Hecht and Tiffany Thayer, was entitled *Doubt*. It carried on the work of the late Charles Fort to record the odd and unusual, though it tried to verify the facts behind the reports. In a 1950 article, they synthesize the discoveries and legends.

This valley, number one legend of the Northlands, has as its background, stories of tropical growth, hot springs, headhunting mountainmen, caves, prehistoric monsters, wailing winds, and lost gold mines. Actual fact certifies hot springs, the wailing winds, and some person or persons who delight in lopping off prospectors' heads. As for the prehistoric monsters, Indians have returned from the Nahanni country with fairly accurate drawings of mastodons burned on rawhide. . . .Already the Indians shunned the place because of its "mammoth grizzlies" and "evil spirits wailing in the canyons."

Canadian police records show that Joe Mulholland of Minnesota, Bill Espler of Winnipeg, Phil Powers and the McLeod brothers of Fort Simpson, Martin Jorgensen, Yukon Fisher, Annie La Ferte, one O'Brien, Edwin Hall, Andy Hays, an unidentified prospector and Ernest Savard have perished in the strange valley since 1910. In 1945 the body of Savard was found in his sleeping bag, head nearly severed from his shoulders. Savard had previously brought rich ore samples out of the Nahanni. In 1946 prospector John Patterson disappeared in the valley. His partner, Frank Henderson, was to have met him there, but never found him.

Mulholland and Espler met their fate in 1936 while O'Brien, Hays, LaFerte and Yukon Fisher perished in macabre circumstances in 1940. In Henderson's case, he came back visibly shaken after searching far and wide for Patterson. He was quoted: "There is no denying the sinister atmosphere of the whole place. The continual wailing of the wind is something I won't forget soon." This wouldn't be the end. In May 1948 Lewis Shebbach, a tough old sourdough gold hunter, left Fort Liard certain he would find the Lost McLeod. It was about a year later when his dried bones were found by Caribou Creek.

The mountain men are described little differently than those strange human *bukwas* or the Dagestan Man. They are, it seems possible, one of those tribes that inhabited Peter Williams, Charley Victor, Chehalis Phillip and J.W. Burns' Saskahaua so long ago.

The last official expedition to have found the lost "Shangri-La Valley" was in August 1931. The leader was Dr. Norman Henry of the Philadelphia Academy of Natural Sciences. Among the expedition were his wife and F.F. McKuster, the latter of the Canadian Topographical Survey. Norman wrote of how they did indeed find a valley with hot springs and a 9 foot pool of mineral water that was 90 degrees in temperature and which they used as a bath. The vegetation was mind-boggling. Wild roses streaked over the hillsides and meadows like a rainbow. Mrs. Henry even later described how "delphiniums grew over 8 feet tall, and raspberries, roses and vetches growing in the thickest, most luxurious tangle I ever saw." Butterflies fluttered everywhere. Exotic birds flew over the valley.

The spellbinding Mackenzies may be the crown and the Saskahaua the navel but in between is the body. This is the wilderness area of the Coast Mountains, the area of the Pacific and Kitimat ranges. This area is etched with fjords and rivers and speckled with lakes. It is a wild and untamed place.

The Rockies form the natural border of eastern British Columbia and are far from explored to the extent of the American Rockies. Near Jasper, famed frontier explorer David Thompson found signs of a 4-toed "mammoth" in 1811, and in 1955 William Roe saw one of the hairy female Sasquatches on nearby Mica Mountain.

But still greater and more daunting is that vast wilderness in between— the Nachako Plateau and the much smaller Omineca Mountains. More than any place Sasquatch and perhaps the Skoocoom live here undisturbed, their chilling mews and howls unheard by man and their kingship over nature unchallenged. Their history is impressive and goes back hundreds of years. It inspired dread and fear. It is a paradox and a mystery— and it is *in Canada.*

Less enticing stories lured the great safaris to Africa, which led to finally conquering the mysteries of the great "Dark Continent." But in our own backyard there is a great unknown and Dark Continent, one that has remained so not because we cannot explore it but because we don't think there is anything to find. It is a fallacy of thought that the exotic must always be far away. We believe it cannot be close at hand. We might find it is closer than we think; at least closer than Africa.

Like wriggling tributaries, mountain ranges branch off from the great Canadian Rockies and lead down into Washington State. Something walks those mountains and avoids the plains. It prefers the cold weather of the far north and of the high altitudes of British Columbia. . .and it prefers to stalk at night. It has little fear of mankind.

It is this network of mountains which are highways for the Sasquatch, and no doubt this explains how it can appear near populated areas in southern British Columbia and perhaps even Washington State, and then disappear for years. It can easily follow these mountains back north to the Rockies and from there to the last range of mountains, the Selwyn and Mackenzies, and between them the Nahanni Valley and that "lost horizon" where blueberries are the size of golf balls, where clusters of grapes burst with juice. . . where possibly nuggets of gold are the size of duck eggs. . .all just south of the forbidding Arctic Circle. It sounds unbelievable only because we have not gone to look. . .again.

This is the land of the Sasquatch. It is green, lush and dense. Rivers and lakes are everywhere. Wild fruit and game are plenteous. It is a vast unknown with which man only flirts. We have too long second-guessed what is in there, and too long profited by mere supposition. This vast Canada and the heights of the great mountains must be challenged.

Edgar Rice Burroughs captivated generations with his novels of isolated lost lands still dominated by "cavemen" and dinosaurs. In the real world we have encountered Dagestan Man and Zana, both in the image of forgotten and innocent imagineering. Legends of the far north tell us of mammoths and a giant salamander and a Shangri-la valley. We must cast aside our cynicism and accept there just really may be a land that time forgot.

Unfortunately, we have seen what we are conditioned to see. We believe what we think must exist based on our own city-bound lives and habits. Much of this is very innocent, but the ignorance it fosters is still ignorance.

Recently a little ditty went through the email, passed on by friends and family and by those of us who found it curious. It was a little test. It asked us to count the number of F's in the following sentence:

FINISHED FILES ARE THE RE-
SULT OF YEARS OF SCIENTIF-
IC STUDY COMBINED WITH THE
EXPERIENCE OF YEARS. . .

If you said three, you are average. Four, you are quite intelligent. If you said six you are a genius! Most people were crestfallen to realize they were only average. There are, in fact, six Fs. Studies had discovered that the English speaking mind cannot process "of." We cannot for one reason: we regard it merely as a preposition and do not give much attention to it. It is a word that merely leads to others, others which contain the gist of what is important. It is right before our face, and a very crucial word, but we have to pay special attention and indeed our minds have to be called to its existence.

In our everyday world we are guilty of the same thing. We take for granted the world of our daily endeavor. We seldom stir from our cities. What's worse, when we do hear of those who have en countered unexplained things we tend to quickly interpret it with our preconceived ideas, ideas based on nothing but communal enforcement and personal experience. We may mock or we may ridicule. But mostly we rationalize and create something entirely false with our imaginations. Those small connective strands are crucial but they are easy to overlook in light of our preconceived ideas, cynicism and paradigms. Yet they tie together all things and make nature logical. When things don't fit and we must rationalize, we know we've overlooked one of those critical prepositions and saw what we wanted.

It is time that we realize that our cities are pockets, and our internet is only wires and wireless signals heading back and forth between them. The world has not shrunk. We have only shrunk *our* world. It is time we wake up to realize that we still live at the threshold of a vast unknown, and in this vast unknown we may just be men amongst the giants.

# BIBLIOGRAPHY

Barnes, Thomas G. D.Sc.  *Origin and Destiny of the Earth's Magnetic Field,* ICR, 1983

Bord, Janet & Colin  *The Bigfoot Casebook,* Stackpole Books, 1982.

Douglas, William O.  *Beyond the High Himalayas,* Doubleday, Garden City, New York, 1952.

Fawcett, Brian, Ed.  *Lost Trails, Lost Cities: An Explorer's Narrative by Colonel Percy Fawcett,* Funk & Wagnalls Co., New York, 1953.

Gish, Duane T., PhD  *Evolution: The Challenge of the Fossil Record,* Creation-Life Publishers, 1985.

Green, John  *On the Track of the Sasquatch: Encounters with Bigfoot from California to Canada .* Cheam Publishing , 1968.

--  --  *Year of the Sasquatch,* Cheam Publishing, 1970, 1971.

--  --  *The Sasquatch File,* Cheam Publishing, 1973.

--  --  *Sasquatch: The Apes Among Us* Hancock House, 1978, 1981.

--  --  *On the Track of the Sasquatch, Book 1,* 1980 Edition, Cheam Publishing, 1980

--  --  *On the Track of the Sasquatch, Book 2,* 1980 Edition, Cheam Publishing.

--  --  *The Best of Sasquatch/Bigfoot,* Hancock House, 2004.

Hillary, Sir Edmund  *High In The Thin Cold Air,* Doubleday 1962.

Hunter, Don,
w/ Dahinden, Rene  *Sasquatch,* McClellend & Stewart, 1973.

Heuvelmans, Bernard, PhD | *On the Track of Unknown Animals*, Hill & Wang, New York, 1958.

Izzard, Ralph, | *The Innocent on Everest*, Hodder & Stoughton, 1954.

--   -- | *The Abominable Snowman Adventure*, Hodder & Stoughton, 1955.

Kane, Paul | *Wanderings of an Artist*, The Radisson Society, Toronto, Canada, 1925.

Krantz, Grover S. *PhD* | *BIG Foot-Prints: A Scientific Inquiry Into The Reality of Sasquatch*, Johnson Book, 1992.

Landsburg, Alan | *In Search of. . .*    Nelson Doubleday, 1978.

LaViolette, Paul A. PhD, | *Earth Under Fire: Humanity's Survival of the Apocalypse*, Starlane Publications 1997.

Long, Greg | *The Making of Bigfoot*, Prometheus Books, 2004.

Morris, Henry M., PhD, | *Scientific Creationism*, Master Books, 1985

Morris, Henry M., PhD,
    Parker, Gary E. EdD, | *What is Creation Science?* Master Books, 1987.

Morris, Henry M., PhD,
    Whitcomb, John C. | *The Genesis Flood*, Presbyterian and Reformed Publishing, 1961.

Napier, John, PhD, | *Bigfoot: The Yeti and Sasquatch in Myth and Reality*, E.P. Dutton & Co., 1973.

Patterson, Roger, | *Do Abominable Snowmen of America Really Exist?* Franklin Press, Yakima, Washington, 1966.

Sanderson, Ivan T. | *Abominable Snowmen: Legend Come to Life*, Chilton 1961, New York

Tchernine, Odette, | *The Yeti,*    Neville Spearman, 1970.

--   -- | *The Snowman and Company*, Robert Hale Lmtd, London, 1961

Tchernine, Odette,  Ed.          *Explorers' and Travelers' Tales,* The
                                 Adventurers Club, 1963

--              -- , Ed.         *Explorers Remember,* The Adventurers Club,
                                 1968.

--              --               *In Pursuit of the Abominable Snowman,*
                                 Taplinger Publishers, 1971, New York.

Slate, B. Ann & Berry, Alan      *Bigfoot,*  Bantam Books, 1976.

Stonor, Charles                  *The Sherpa and the Snowman,*  Hollis &
                                 Carter,  London, 1955.

Shackley, Myra                   *Wildmen,*    Thames & Hudson, 1983

Tilman, H.W.                     *Mount Everest 1938,* Cambridge University
                                 Press, 1948, Cambridge, England.

--      --                       *Nepal Himalaya,* Cambridge University
                                 Press, 1952.

Wilkins,  Harold  T              *Mysteries of Ancient South America,* Citadel
                                 Press, New York, 1956.